On the Air with Ham Radio

Your guide to the fascinating ways hams communicate!

Steve Ford, WB8IMY

ARRL *The national association for* **AMATEUR RADIO**

225 Main Street • Newington, CT 06111-1494
ARRLWeb: **www.arrl.org/**

The Cover

Top: Bruce Steele, WØVET, enjoys the view—and a QSO—from Scotts Bluff National Monument in Nebraska.

Center left: There's nothing quite like Field Day, as Ken, W3USN, demonstrates.

Center right: Young William Wynn, KF6ZFR, enjoys Kid's Day, an ARRL-sponsored event that gives young people (licensed or not) the chance to make new friends on the air. Dad N6AXD is at the left, and granddad Larry, W6AXD, is at the right.

Bottom: Graphic by David Pingree, N1NAS.

Back Cover: Slow-scan TV image of the smiling face of K8JGY.

Contents

Foreword

If you're a new ham, you've been patient for a long time. You've struggled through studies and examinations, only to wait while the FCC processed your documents and awarded that precious Amateur Radio call sign.

But now you finally have it and you're itching to put the license to work. Suddenly, a flurry of questions arises out of nowhere…

• Do I want to explore the world of VHF and above, or would I be more comfortable starting on the HF bands?

• Which transceiver should I buy? What about an antenna?

• Which mode should I use? SSB, CW, video or any one of the many digital modes?

• How do I hook up all this hardware?

• How do I make contacts with my chosen mode?

If you're lucky, you may have an *Elmer* to help you. In Amateur Radio tradition, an Elmer is a knowledgeable ham who can answer all your questions and help you avoid serious pitfalls as you begin your journey. But if you don't have the luxury of an Elmer (many amateurs these days do not), what can you do?

Think of this book as your "paper Elmer." It doesn't contain every fact in the Amateur Radio universe, but it contains the most important information you'll need to get started on the right foot. The reading is easy, and often humorous. Each chapter is devoted to a particular type of Amateur Radio experience. As you breeze through the book, you'll take a guided tour of much that ham radio has to offer. Pick a facet that interests you and you'll have all the practical advice you need at your fingertips. This is the kind of how-to book that never goes out of style.

Welcome to Amateur Radio!

David Sumner, K1ZZ
Executive Vice President
Newington, Connecticut
March 2001

About the ARRL

The seed for Amateur Radio was planted in the 1890s, when Guglielmo Marconi began his experiments in wireless telegraphy. Soon he was joined by dozens, then hundreds, of others who were enthusiastic about sending and receiving messages through the air—some with a commercial interest, but others solely out of a love for this new communications medium. The United States government began licensing Amateur Radio operators in 1912.

By 1914, there were thousands of Amateur Radio operators—hams—in the United States. Hiram Percy Maxim, a leading Hartford, Connecticut, inventor and industrialist saw the need for an organization to band together this fledgling group of radio experimenters. In May 1914 he founded the American Radio Relay League (ARRL) to meet that need.

Today ARRL, with approximately 170,000 members, is the largest organization of radio amateurs in the United States. The ARRL is a not-for-profit organization that:

- promotes interest in Amateur Radio communications and experimentation
- represents US radio amateurs in legislative matters, and
- maintains fraternalism and a high standard of conduct among Amateur Radio operators.

At ARRL headquarters in the Hartford suburb of Newington, the staff helps serve the needs of members. ARRL is also International Secretariat for the International Amateur Radio Union, which is made up of similar societies in 150 countries around the world.

ARRL publishes the monthly journal *QST*, as well as newsletters and many publications covering all aspects of Amateur Radio. Its headquarters station, W1AW, transmits bulletins of interest to radio amateurs and Morse code practice sessions. The ARRL also coordinates an extensive field organization, which includes volunteers who provide technical information for radio amateurs and public-service activities. In addition, ARRL represents US amateurs with the Federal Communications Commission and other government agencies in the US and abroad.

Membership in ARRL means much more than receiving *QST* each month. In addition to the services already described, ARRL offers membership services on a personal level, such as the ARRL Volunteer Examiner Coordinator Program and a QSL bureau.

Full ARRL membership (available only to licensed radio amateurs) gives you a voice in how the affairs of the organization are governed. ARRL policy is set by a Board of Directors (one from each of 15 Divisions). Each year, one-third of the ARRL Board of Directors stands for election by the full members they represent. The day-to-day operation of ARRL HQ is managed by an Executive Vice President and his staff.

No matter what aspect of Amateur Radio attracts you, ARRL membership is relevant and important. There would be no Amateur Radio as we know it today were it not for the ARRL. We would be happy to welcome you as a member! (An Amateur Radio license is not required for Associate Membership.) For more information about ARRL and answers to any questions you may have about Amateur Radio, write or call:

ARRL—The national association for Amateur Radio
225 Main Street
Newington CT 06111-1494
Voice: 860-594-0200
Fax: 860-594-0259
E-mail: **hq@arrl.org**
Internet: **www.arrl.org/**

Prospective new amateurs call (toll-free):
800-32-NEW HAM (800-326-3942)
You can also contact us via e-mail at **newham@arrl.org**
or check out *ARRLWeb* at **www.arrl.org/**

1 The Spirit of Radio

They call our hobby Amateur Radio. The word "amateur" comes from the Latin "amator"—*lover*. We're lovers of radio. We're enamored of the notion of communicating over great distances without any physical connection.

Of course, the nonbelievers don't understand our obsession.

"Couldn't you just pick up the telephone and *call* the people you're talking to?" Blasphemy!

They don't "get it" and probably never will. They can't appreciate the magic of sending and receiving signals with your bare hands, talking to people throughout the world using little more than a wire strung between two trees. They only know communication as it exists through the graces of multibillion dollar corporate and government networks. They grab a telephone, punch in a number and talk. How their voices get from here to there is of little consequence to them. To us, it's *everything*!

But why do they call us "hams"? Is it because we're weird and obnoxious? Actually, the history behind the word "ham" is a little fuzzy.

"Ham" was not a complimentary term when it was first applied to our hobby. According to G. M. Dodge's *The Telegraph Instructor*, published well before the advent of radio, a "ham" was a "bad operator." As landline telegraphers became wireless telegraphers, they carried this terminology with them.

In the early days of radio, most stations communicated by generating a spark of electricity between two electrodes. *Spark-gap* transmissions were notoriously "dirty" and occupied large chunks of the radio spectrum. In the unregulated environment that prevailed, interference was a huge problem. Government stations, ships, coastal stations and Amateur Radio operators all slugged it out for signal supremacy in each other's receivers. Amateur stations were often very powerful, and two hams working each other across town could effectively jam all the other operations in the area. This caused the commercial guys to complain bitterly on the air about, ". . . THOSE #$!@ HAMS JAMMING EVERYTHING."

Amateurs, possibly unfamiliar with the real meaning of the term, picked it up and applied it to themselves. As the years passed, the original derogatory meaning disappeared. To this day, the nonbelievers may not know us as Amateur Radio operators, but say "hams" and you'll probably see a glimmer of understanding in their eyes.

"Oh, yes. It's that radio *hobby*."

No, Amateur Radio is much more. It's also the spirit of service to the community. When a tornado rips through a town and destroys all conventional means of communications, you don't hear the authorities say, "Quick! Call out the model train collectors!" In times of crisis, hams are among the first to jump into action. Not only do we have the radios, we know how to use them. We're trained communicators.

If you ever visit ARRL Headquarters in Newington, Connecticut, look for a small, granite monument that rests about 100 feet to the left of the main building entrance. On that stone you'll find the names of hams who have given their lives while using their *hobby* to save others. Do you think a similar memorial exists for bird watchers?

As a ham you'll talk to people in the most far-flung places imaginable. Marek, SP3GVX, is the chief radio operator at the Polish Antarctic Research Station on the South Shetland Islands. When he isn't involved in his professional duties, Marek enjoys Amateur Radio from this station by signing HF0POL.

IS HAM RADIO EXPENSIVE?

When you're taking your first steps into a new venture, it's wise to keep your eyes wide open. Unfortunately, many Amateur Radio books and articles try to minimize the issue of cost, or avoid it altogether. The authors have the best intentions. They're worried about scaring you away from the hobby.

I'd like to take a different approach with this book. I believe that you're mature enough to understand that there can be some costs involved. On the other hand, Amateur Radio is not the most expensive hobby in the world. If you want to see how bad it can *really* get, talk to a private pilot who owns his or her own airplane. General aviation makes hamming look like rock collecting by comparison.

Randy, AL7PJ, operates from Alaska's Matanuska Glacier.

There is a decent chance that you'll fork over at least $200 to get started in Amateur Radio, probably more. Two hundred dollars is a lot of money for many people. (It annoys me how some authors attempt to trivialize the impact of "a few hundred dollars.")

You'll probably hear that you can save huge amounts of money by building your own gear from kits. This is true to a degree. Most transceiver kits are low-power HF rigs, although you'll find the occasional VHF or UHF kit. If you have sufficient technical experience to build a

Steve, WB4OMM, enjoys on-the-air contesting—particularly the ARRL November Sweepstakes.

kit and fix it if something goes wrong, a kit is a viable cash-saving option.

But if you want a transceiver with sophisticated features and higher power output, and if your engineering skills are lacking, you're looking at new or used equipment. Now we're talking about potentially large price tags.

The good news is that you won't be shelling out the big bucks on a frequent basis. A new dual-band FM mobile transceiver can set you back about $500, but you probably won't buy another one for a *long* time. If you take good care of a transceiver, it can theoretically last a lifetime. There are hams today who are using transceivers manufactured in the 1950s. Those radios work just as well today as they did when they were new.

Set reasonable goals for yourself and you can save even more money. For example, forget about tall towers and huge antennas for now. They're *very* expensive. Instead, fix your sights on less ambitious antennas that you can install on your roof, patio or wherever. If your passion is to operate on the HF bands, consider wire antennas. You'll have a blast with a wire *dipole*, at a cost of about $25 or less.

When you see those photographs of elaborate, wall-to-wall ham stations and forests of steel towers, consider two things:

(1) **The ham in question might be rich.** There's nothing wrong with wealth, but most of us don't have that advantage at the moment.

(2) **The gear may have been accumulated over a long period of time**. This is what usually happens. You buy a radio. A couple of months or years later, you buy another. Perhaps you receive a couple of accessories as gifts. After a decade of purchasing and gift giving, you wind up with a station that would make NASA jealous!

The Boy Scouts of America Venture Crew 828 enjoy low-power (QRP) hamming.

The Nerd Factor

Since the earliest days of Amateur Radio, we've been tagged with an annoying stereotype: *nerds*. What's a nerd? Putting it simply, a nerd is someone of high intelligence, poor social skills and even poorer personal hygiene. The modern nerd wears clothes that are roughly 20 years behind current fashion. His hair is matted and rarely washed. Caucasian nerds tend to be quite pale in color, due to high amounts of time spent indoors. Nerds of all races tend to speak in short, clipped sentences. Their speech is occasionally accented by various facial twitches; a nerd rarely looks you in the eye when speaking. If the conversation departs from ham radio, Star Trek, the Hobbit, Dr Who, or Monty Python, the nerd usually falls silent.

Does this sound like anyone you know?

Probably not. That's the nature of stereotypes. They paint a picture with a brush that's a mile wide. I confess that I've met some hams who fit the previous description perfectly, but they are the exceptions, not the rule. It's true that Amateur Radio attracts people who are interested in science and technology. Some of these individuals are indeed "nerdish," but it's usually a byproduct of extremely high intelligence. The nerds I've met are very focused, intense people. They are completely immersed in technology, and Amateur Radio gives them an avenue to experiment and stretch their horizons.

But the majority of amateurs are as normal and well adjusted as anyone else. For most of us, Amateur Radio is not an all-consuming passion. No matter how much we love the magic of radio, our academic studies, our jobs, our friends and our families come *first*.

THE SOCIAL SIDE OF HAM RADIO

Hamming is much more than hardware. In the days before political correctness expunged sexist language from our vocabularies, Amateur Radio was called a "fraternity." Although the word has fallen out of fashion, the idea is still valid. There is a bond between hams that defies an easy description. Taking my best shot, I'd say the bond is similar to what you'd find among members of organizations such as the Freemasons, Knights of Columbus, Odd Fellows and so on. It's the idea that we hams share "secret" knowledge that is beyond the understanding of the general public. And, to a great extent, that's true! When you consider the fact that most people can't set the digital clocks on their VCRs, they must think we're magicians!

That fraternal bond manifests itself in many ways. You might be talking to someone at a party and casually mention that you're a ham. "Really? So am I! I'm KD7XZY!" You give him your call sign and it's as though you've just exchanged a secret handshake.

Perhaps you're on vacation and you've just accessed a repeater in a strange

Matti Rouhiainen, OH2PO, in Finland has built these impressive antennas by hand!

town. You ask for directions to a restaurant. Someone answers your call, gives you the directions . . . then asks if you'd like to meet for breakfast the following morning!

Like many hams, I used to have my call sign on the license plates of my car. Once I was driving through eastern Montana. There was nothing to see but rolling fields of wheat. I seemed to be the only traveler on the road. Soon another car appeared behind me. As it drew closer, the driver began to honk. I had no idea what was going on until, to my astonishment, I realized that he was honking in Morse code, sending my call sign over and over!

For the next 20 minutes we cruised down the interstate, honking to each other like lunatics. He asked if I had a 2-meter FM radio. No, I replied. So he sent his call sign, told me his name was Ralph, and said that he was on his way to Miles City, Montana. I pounded the horn switch, sending my particulars in sloppy CW. After a while he said that he had to exit for gas and ended with a snappy "73" (best wishes). That was, without a doubt, the most unusual CW conversation I've ever enjoyed! Can you think of any other hobby that would inspire a scene like that?

HAMFESTS

When hams feel the need to meet each other face-to-face, they often do it at gatherings known as *hamfests*. Even the smallest hamfests usually include a flea market where hams sell and trade used equipment. If you're in the mood to spend money on preowned gear, a hamfest is the place to do it. The larger hamfests will also attract new-equipment dealers, and the *very* largest will have exhibits from the manufacturers themselves.

Medium to large-size hamfests will often offer forums where you can take part in discussions on various topics. Some forum topics are technical while others are political. Many hamfests also have food available on the premises, and even entertainment.

No matter where you live, there is probably at least one hamfest each year

More than 28,000 amateurs show up for the annual Hamvention in Dayton, Ohio.

Ken, W3USN, makes a late-night CW contact from a tent during Field Day weekend.

within 100 miles of your location. If you live in the more populated regions of our country, you can count on having a dozen hamfests or more within a comfortable driving distance. Check the Hamfest Calendar in each issue of *QST* magazine or on the ARRLWeb site (**www.arrl.org/**) to see what's coming up in your area. Your local club may sponsor a hamfest. If not, it's a safe bet that someone in the organization knows when the nearest hamfest takes place.

But just as Moslems are urged to make at least one pilgrimage to the holy city of Mecca, you must make at least one journey to the largest hamfest on the face of the planet: the Dayton *Hamvention*. Each year in late May, 25,000 Amateur Radio operators swarm into Dayton, Ohio, and converge on the Hara Arena convention center. The result is a gigantic ham radio flea market, along with the most extensive collection of Amateur Radio dealers and manufacturers in the world. If you can't find it at the Dayton Hamvention, it may not exist!

The three-day event draws a vast cross section of the human species. Some make the pilgrimage to sell, others to buy. They march up and down the endless aisles of the outdoor flea market and cram themselves into the arena complex. The air is so full of radio signals (mostly from FM hand-held transceivers), interference is horrendous. Cars honk unceasingly. Television news helicopters hover overhead. The odor of hot dogs, beer, cigarettes and soft drinks wafts through the hallways. If you're lucky, it won't rain . . . or at least the tow trucks will be able to rescue your car from the mud if it does.

Why would anyone subject themselves to such a spectacle? It's that darn fraternal notion again. As you shuffle wearily through the crowds you can't help catching the glances of equally weary souls. That millisecond meeting of the eyes says it all: "Greetings, fellow magician. We're suffering, but we're here. Isn't it grand?"

DON'T SIT STILL

Books are great things, but don't keep your nose buried in one too long! You won't savor the spirit of radio by flipping pages.

William Wynn, KF6ZFR, thoroughly enjoys Kid's Day on the air—when he can wrestle the radios away from his grandfather Larry, W6AXD (right) and his father David, N6AXD (left).

This book is *not* the be-all end-all reference for Amateur Radio. You won't, for example, find construction projects in this book. Pick up a copy of *The ARRL Handbook* or *The ARRL Antenna Book* and you'll find a lifetime's worth of projects. And if you need more detail on every Amateur Radio communication mode and how to use it, buy *The ARRL Operating Manual*. These books are available from your favorite dealer or directly from ARRL Headquarters.

The purpose of this tome is to get you on the air as quickly and painlessly as possible. You'll learn which hardware combinations work best, and how to put them together to create a station for a particular mode. You'll also receive valuable operating information to help you avoid the "newbie" label!

Browse the chapters and decide which activities interest you the most. (The chapters are organized according to the most popular Amateur Radio pastimes.) Then, read your chosen chapters carefully. You'll find that the reading is easy and even fun.

After that, *put the book down*, get on the air and start enjoying Amateur Radio! It's okay to reread the book to refresh your memory, or when it's time to explore a new facet of our hobby, but there's no substitute for the total joy of communicating without wires. The sooner you start burning the air with electromagnetic waves of your own creation, the sooner you'll appreciate the magic!

2 | FM—No Static at All

When you hear the word "FM" you probably think of a station that blares the hits somewhere between 88 and 108 on your radio dial. The music is crystal clear and there are few, if any, annoying buzzes or other noises.

But FM actually stands for *frequency modulation*. It's a method of transmitting information that involves shifting the frequency of a radio signal back and forth in sync with voices, music or whatever. (The amount of frequency shift is known as *deviation*.)

AM means *amplitude modulation*. It's almost the reverse of FM. An AM transmission is comprised of a *carrier* signal and two *sidebands*. You send information on AM by shifting the *strength* (amplitude) of the sidebands. Unlike FM, however, the frequency of an AM signal never changes.

If you built a radio that listened *only* for signals that shifted their frequencies, you wouldn't hear AM signals at all, would you? And since most of the static in the world is amplitude modulated (Mother Nature must be fond of AM), your clever radio would automatically reject noise! Congratulations. You've just designed an FM receiver. (You're about 75 years too late, but who's counting?)

As you've probably guessed by now, FM transmissions can take place on *any* frequency, not just those between 88 and 108 MHz. It's federal law, courtesy of our friends at the Federal Communications Commission (FCC), that dictates where FM signals can appear. Hams are allowed to transmit FM mainly on our VHF and UHF frequencies (those above 50 MHz). We also have a small FM segment between 29.5 and 30 MHz.

WHY FM?

When it comes to clear two-way communication, FM is way ahead of AM. It's a pleasure to cruise the highways and chat with your buddies without noise interference. You can be driving through the granddaddy of all thunderstorms and hardly hear a peep of lightning static in your radio. Try the same thing with an AM transceiver and you'd end up deaf, insane or both.

Assuming that your radio is connected to a decent speaker, FM audio has wonderful fidelity. If your friend is speaking clearly (as opposed to screaming and cursing), you'll hear every word. During some conversations it sounds as though the person is right there in the car with you—whether you like it or not.

The only skunk at the party is *range*. Unlike shortwave signals, VHF and UHF transmissions rarely go bouncing off the upper layers of the atmosphere to land on other continents. Instead, they take the straight-line express route to outer space. This means that your communication range is limited to a short distance beyond your local horizon *at best*. (Yes, there are exceptions. We'll talk about those in later chapters.)

To make matters worse, an FM receiver requires a reasonably strong signal to produce all that nice, clean audio. When an FM signal becomes weak, the receiver doesn't do a very good job of producing something you can understand.

So, to cover a worthwhile amount of territory on VHF and UHF, you need a little help.

ENTER THE FM REPEATER

In simplest terms, a repeater *repeats*. It's an electronic parrot. A repeater listens to a signal and repeats what it hears. Unlike a parrot, however, a repeater repeats the signal *while it's listening to it*.

From a technical point of view, a repeater is a station that operates automatically. There is an antenna, a transmitter and a receiver, but no human operator pressing buttons and flipping switches. A microprocessor-based device known as a *controller* is the brains of the outfit. The controller makes sure everything operates properly, or takes the repeater off the air if it doesn't.

The repeater's receiver is exquisitely sensitive, and its transmitter usually operates at high power, typically 100 W or more. The antenna is an *omni-directional* type. (In other words, it receives and transmits in all directions.) If you place a repeater system on a mountaintop, hill, skyscraper or radio tower, it can receive and transmit over an enormous distance—even on VHF or UHF.

Take a look at **Figure 2-1**. Let's say your puny mobile radio can only cover a few miles on the ground. You don't have a snowball's chance of contacting someone, say, 20 miles away. But the repeater, from its lofty perch, hears both you *and* the other fellow.

The instant you begin speaking, the repeater receives your signal and blasts it out over hundreds or thousands of square miles. You keep talking and it keeps repeating. Thanks to the repeater, your distant buddy hears your melodious voice booming in his radio. Of course, he isn't picking up your signal directly. He's listening to the transmissions of the repeater.

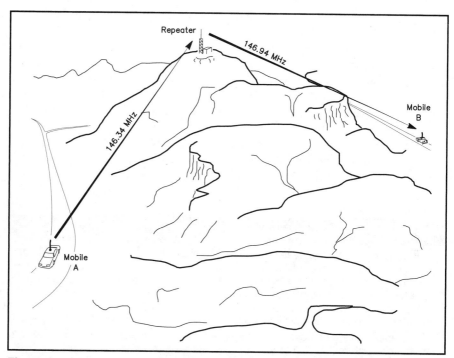

Figure 2-1—The mountain blocks direct communication between mobile stations A and B. When mobile A transmits, however, the repeater hears its signals and relays it to mobile B.

Only one person can use a repeater at a time. When you release your microphone button, it's your friend's turn to talk. The repeater picks up his signal and relays it back to you. Once again, you're not hearing him directly. You're listening to the repeater acting as the middleman in your conversation.

One of the keys to this technological miracle is the repeater's ability to receive and transmit simultaneously. This may seem like a simple trick, but it isn't.

Separate Frequencies, Separate Energies

Have you ever watched someone use a public address system that's horribly out of adjustment? Some poor soul strolls up to the microphone, clears his throat and whispers, "Is this thing on?"

The amps are usually turned up so high that the sound of his amplified voice is picked up by the microphone. The PA system quickly amplifies the voice again, and again, and again until a banshee howl fills the auditorium and sends everyone diving under their seats. This painful phenomenon is known as *feedback*.

Feedback can happen in radio devices just as easily, and the consequences can be just as awful. If a repeater transmitted and received simultaneously on the same frequency, you'd have a horrendous feedback loop. Yes, the repeater would howl, too!

So what's a poor repeater to do? The answer is to transmit and receive on *separate* frequencies. Let's say we have a repeater that listens for signals on 147.96 MHz and repeats whatever it hears on 147.36 MHz (see **Figure 2-2**). Notice that the input and output frequencies are separated by 600 kHz. This is the standard for 2-meter repeaters, although you'll occasionally find repeaters using different *splits* or *offsets*.

On 146 MHz the standard offset is 600 kHz. One MHz is the common split on 222 MHz and a 1.6-MHz separation is the plan at 440 MHz. There are exceptions to this rule, as experience will teach you.

But just when you thought all these heady concepts were falling into place,

Who Builds Repeaters?

Building a repeater system is an expensive proposition. Even the most basic repeaters can cost several thousand dollars. The money isn't all tied up in equipment, either. There is often a rental charge for the building that houses the electronics. (Those mountaintop shelters can be expensive bits of real estate!) In some cases there are fees just for having your antenna clamped to a commercial tower. If the repeater system has a telephone interface, there is a monthly invoice for that, too. And don't forget the electric bill!

Repeaters are so costly that most are installed and operated by clubs. If you have 100 members contributing $10 or $20 apiece, that's usually sufficient to maintain all but the most complex systems. Clubs are often able to get breaks; perhaps free use of a building or tower, for example. Most clubs prefer that you become a member if you intend to use their repeaters on a regular basis. By signing on as a member you may receive "secret" codes that allow you to operate some of the repeater's special functions—such as the autopatch.

Can Anyone Put a Repeater on the Air?

Back in the gold rush days of amateur FM, repeaters popped on the air like mushrooms after a summer rain. The atmosphere was totally Dodge City—little law and even less order. But as repeaters became more popular, they began bumping into each other. Interference became the rule, not the exception, in many areas. That's when *repeater coordinating groups* began forming throughout the country.

Coordinating groups are simply hams who work together to keep the chaos to a minimum. Unless a club intends to put a renegade repeater on the air—one that could draw unwelcome attention from the FCC—they must first consult with the coordinators. The coordinators will tell them which frequencies are available, where potential interference problems exist and so on.

The 2-meter amateur frequency band is filled to capacity in most areas. That's why you'll often see new repeater systems appearing on 222 and 440 MHz, if not beyond. Thanks to the coordinators, repeater owners manage to share the available space without coming to blows.

Their greatest headaches these days are caused by dual-band VHF/UHF radios that have the ability to function as crossband repeaters. They can transmit a 440-MHz signal on 2 meters, or vice versa. Some hams are throwing them on the air without considering the possible conflicts. If you own one of these electronic wonders, *think* before using the crossband feature. The best advice is to use it sparingly, if at all.

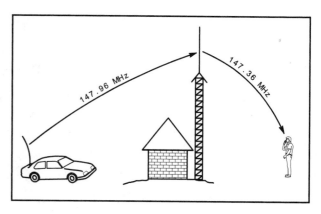

Figure 2-2—Repeaters use a split-frequency scheme to avoid nasty feedback problems. In this case, the repeater receives the mobile operator on 147.96 MHz and transmits what he is saying on 147.36 MHz.

here's a new curve ball for you. Not only do repeaters transmit and receive at the same time, most do it *on the same antenna!*

Say *what*?

You heard right. Crazy as it seems, there's no law against receiving and transmitting simultaneously with the same antenna. The incoming and outgoing waves of energy pass like ghosts in the night. They don't smash into each other like a cosmic freeway pileup.

The trick, however, is keeping the high-power transmit energy from finding its way into the sensitive repeater receiver. A repeater receiver is designed to deal with signals carrying the energies of a butterfly sneeze. If the repeater transmitter was allowed to dump even a fraction of its power into the receiver, the result would be the equivalent of dropping a 50-pound cinder block on a Sony Walkman. Not only would the receiver be deaf as a post, its life span would be measured in milliseconds.

The magical device that keeps the receiver from being cooked to a golden brown is known as a *duplexer*. Duplexers are extremely sharp filters that keep the transmit and receive signals separated at the point where they enter or exit the antenna system. They are impressive sights, as you can see in **Figure 2-3**. They look like big metal cylinders. Duplexers for 2-meter repeaters can be more than three feet long. Six-meter repeater duplexers are true monsters with some as tall as six feet or more.

Be careful not to confuse a duplexer with a *diplexer*. Diplexers (and *triplexers*) are used to separate transmit and receive signals when they're at vastly different frequencies (usually different bands altogether). They're compact devices, intended for use with dual- and tri-band VHF/UHF transceivers. They allow you to operate these rigs on multiple bands with a single antenna.

When it comes to separating signals that are much closer in frequency, only a duplexer will do. Unfortunately, you'll often see diplexers referred to as duplexers in advertisements and catalogs. The difference is easy to spot, though. True duplexers cost $1000 or more, and you can't fit them under your dashboard!

Figure 2-3—These strange-looking cylinders are duplexers. They keep the received signals and transmitted energy from mixing in destructive ways!

FM Without Repeaters

We've clearly established the notion that repeaters are wondrous devices. They make it possible for you to talk over a humongous area with just a watt or two from your radio. But the equation **FM=Repeaters** is *not* written in stone!

If your signal will span the distance on its own, why use a repeater at all? You can talk from station to station—from your antenna directly to his—and enjoy much more privacy than you do on a repeater. And you don't need to worry about hogging the repeater with your endless observations on the future of mankind, the hottest gossip, or whatever.

Direct communication on a single frequency is known as *simplex*. It simply means that you're communicating in one direction—and one direction *only*—at any given time. *Duplex*, by comparison, means that you're communicating in both directions simultaneously. Telephones are duplex devices because you can speak and listen at the same time.

There are FM simplex frequencies set aside on every amateur VHF and UHF band. If you're looking for random contacts, however, keep your ears tuned to the national FM simplex calling frequencies. They're shown in the sidebar, "Where's The Action?" Of all the national calling frequencies, the most popular is 146.52 MHz. If you hear someone say, "Let's go simplex," or "Let's go to 5-2," they usually mean 146.52 MHz.

If you live in or near a large urban area, don't use the national simplex calling frequencies for long conversations. There's a decent chance that you'll run into interference from other stations.

The rule of thumb is to use simplex whenever possible. Keep the repeater free for hams who really need it. If you can talk to the other station without the repeater's help, why not let someone else use it?

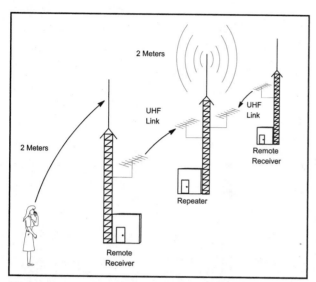

2 Meters

UHF Link

UHF Link

2 Meters

Remote Receiver

Repeater

Remote Receiver

Figure 2-4—Many repeater systems use more than one receiver to provide reliable coverage over a wide area.

Let's Get Complicated

So far we've been talking about simple repeater systems; one transmitter, one receiver and one antenna. Repeaters can become much more complicated, however.

Figure 2-4 illustrates a receiver system that has one transmitter, but many receivers. This is a great way to fill the "holes" in a repeater's coverage. Each receiver sends whatever it hears back to the main site via a UHF or microwave link. At the site there is a device that analyzes all the incoming signals and selects the strongest one to pass to the repeater transmitter.

You can usually tell when you're listening to a repeater system that's using remote receivers. If someone is carrying on a conversation while driving around town, they'll pass in and out of the coverage area of various receivers. As they do, you'll hear their signal become noisy, then switch to crystal clarity. (Analog cell phone users will recognize this!)

Repeater complexity doesn't end with multiple receivers. A repeater can have multiple receivers and transmitters *on other bands*. This allows *crossband linking*. For example, a 2-meter repeater may have a receiver and transmitter on the 222-MHz band. The same repeater may have a transmitter and receiver on the 10-meter FM subband. When 10 meters opens up, everyone on the system—including the folks on 222 MHz—can jump into the fray. (Before you ask, yes, this is all perfectly legal.)

The ultimate repeater coverage scheme is the *linked* system (see **Figure 2-5**). In this type of system, many repeaters are connected through UHF or microwave links to create a huge network that can spread over hundreds or thousands of miles. On a linked system, a ham in one city can talk to other hams in distant cities—with nothing more powerful than a hand-held transceiver.

Linked repeater systems are popular in the Midwest and Far West. The Evergreen Intertie, for example, covers much of Oregon, Washington, and even parts of British Columbia.

YOUR OWN FM RADIO

In the amateur FM universe a "station" can be a tiny hand-held transceiver

Figure 2-5—Linked repeater systems allow FM-active hams to cover huge areas—sometimes several states at once—with nothing more than an H-T.

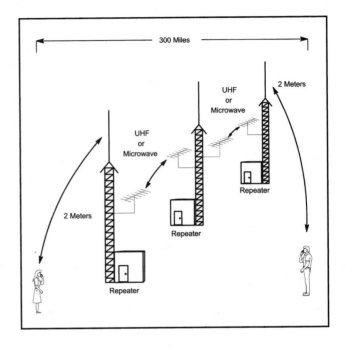

you wear on your belt. It can also be an electronic assemblage that occupies half your writing desk. The choice is yours, depending on your cash flow and the temperament of your spouse, parents or whomever.

H-Ts

The hand-held transceiver, or *H-T*, is the trademark of the FM operator. If you go to a hamfest you'll see them everywhere. They dangle from belts like six-shooters in an Old West movie, although I've never witnessed two hams engaging in an H-T shoot-out. Many H-Ts also sport speaker/microphones and these are clipped to lapels and shirt pockets.

Of course, an H-T wouldn't be a true radio without the ubiquitous *rubber duck* antenna. If you're old enough to remember the CB craze of the mid-'70s, you no doubt recall C. W. McCall's languid drawl, "This here's the Rubber Duck, come on." Well . . . this rubber duck has nothing to do with that god-awful song! (Not all music memories are good!) The rubber reference refers to the fact that the antenna is encased in flexible plastic or rubber. The duck part just comes naturally to some folks.

H-Ts are wonders of miniaturization. The manufacturers pack everything but the proverbial kitchen sink into these radios. Your typical H-T puts out about 2 W of RF power, enough to pack a solid signal into a sensitive repeater. Some will produce even more power with optional battery packs. *DTMF* (*Dual-Tone Multi Frequency*) telephone-style keypads are standard equipment, as are bright liquid-crystal displays (LCDs). Most H-Ts also offer extended receive capability (listen to more than just hams!), paging (bug your buddies anytime, anywhere), scanning (never miss a single signal) and much more.

The downside of all these goodies is that modern H-Ts are notoriously complicated. Learning to simply program a repeater channel could require a substantial chunk of time. If you want to delve into the netherworld of paging or other arcane magic, set aside a millennium or two. (Just kidding. A decade ought to do the trick.)

H-Ts are available in single-band models, dual banders and—for the ultimate in mind-bending complexity—*tri*banders. Obviously, the more bands the radio offers, the deeper you'll need to dig into your checking account. If most of the activity in your area is on the 2-meter band, save your cash and buy a single-band radio. If you live in a densely populated region, you're more likely to have large numbers of repeaters on more than one band. In this situation a dual-band H-T makes sense. Even a tribander might be worth the investment in some areas. Before you reach for your wallet, check *The ARRL Repeater Directory* and determine what bands are most active where you live. Don't buy more radio than you need!

So, should you run out and purchase your own H-T? Here are a few guidelines . . .

Buy an H-T if:
- You plan to operate primarily in your car, on foot, or on a bicycle. An H-T's portability is terrific for hams on the go.
- You have at least one powerful repeater in your area. With the low output power of an H-T, you need a repeater that can relay your signals reliably.
- You plan on doing a fair amount of public-service work. H-Ts are almost mandatory for public-service activities these days. When you're providing communication for a foot race, for example, you don't want to be tied down to your car.

Base/Mobile Transceivers

Like H-Ts, mobile transceivers have benefited from the revolution in miniaturization. The typical FM mobile radio is probably smaller than the AM/FM broadcast radio in your dashboard. Even so, most pack a big RF punch with output power ranging from 30 to 50 W. That's more than enough "oomph" to hit all but the most distant repeaters.

Mobile rigs are also packed with every feature imaginable. A mobile radio can usually do anything an H-T can, and more! Despite their small sizes, mobile rigs are easier to use than H-Ts. The buttons and other controls are farther apart, making it possible for even the chubbiest fingers to use them. The LCD displays are also larger. A larger display is much easier to read.

Generally speaking, the receive audio quality of a mobile radio is superior to that of an H-T. Mobile speakers are larger and the case offers at least some acoustic baffling. Audio clarity and power are very important for mobile operating. When the wind is whistling through the open window and your favorite song is blaring at the threshold of pain, you still want to be able to hear your radio, don't you?

A mobile radio installation is clean and convenient. (Assuming you don't go after your dashboard with a chain saw and a belt sander.) You can install the radio in the dash, or under it. The antenna cable connects to the back panel, out of sight. The same is true of the dc power cable. Many mobile transceivers are designed to mount on rails that allow you to slide the rig out and take it with you when you leave the car. A few can even be installed in the trunk with just a tiny control panel and microphone connector in your dashboard. (Imagine the consternation of the thief who steals that control panel. "Gee, this dang radio don't work! I wonder why?")

Mobile rigs can also do double duty as *base* radios in homes, apartments, offices or wherever. When a mobile transceiver is in your car, it's drawing power from the automotive battery. In your home, you'll need to provide the same dc power. You do this through a device known as a *power supply*. A power supply takes the 120 V ac from the wall outlet and converts it to 12 V dc for the radio.

There's No Such Thing as Too Much Current

The key to buying a power supply is getting the right *current capacity*. Your mobile transceiver draws the highest amount of current when you're transmitting. So, the power supply must be capable of providing *at least* that much current on a *continuous basis*. Beware of ads for high-current power supplies that seem incredibly inexpensive. Read the specifications. If the power supply is rated at 20 A, is that continuous or intermittent (often called "ICS" or "surge")? The intermittent current rating will always be much higher than the continuous rating. If your radio needs 20 A of current when you're transmitting, your power supply must be able to supply 20 A *continuously*. Don't worry about buying a power supply that offers more current than you need. Your radio will draw only as much as it requires. If you can find a good deal on a power supply that offers two or three times as much juice as you need, go for it! You can always use the extra current capacity to power other goodies in the future.

Used or New?

Unless you're a gambler, avoid *large* investments in used equipment—radios or otherwise. The exception is when you can buy your gear from a reputable dealer who'll stand behind it for at least 30 days. Sometimes it takes that long for problems to show up.

If you're a whiz at fixing electronic circuitry, or know a good friend who is, you're in a different category. There are plenty of used-equipment bargains out there. Check out your nearest hamfest flea market, or online auction sites on the Web such as eBay (**www.ebay.com**), and I bet you'll find plenty of used FM radios for very reasonable prices. Some of these radios are great. You fire them up and they perform like new rigs. Others are . . . well . . . disappointing.

Hams have excellent reputations as honest peddlers of second-hand gear. Even so, there are exceptions to the rule. I've been told that something was working "just fine," only to discover that the seller was exaggerating quite a bit! Once I had to throw away a large item that I'd purchased at a hamfest. After trying to make it work, I discovered that it was damaged beyond the cost of fixing. I've also put in some very long hours repairing other "bargains."

If you have any doubt about your technical abilities, stick with new radios. You'll enjoy the advantage of the manufacturer's warranty. If the rig gets sick during the warranty period, it's not your problem.

ANTENNAS

The best radio in the world isn't worth a stale donut without a good antenna. Imagine a stereo system with a $3000 amplifier and $20 speakers. How do you think it will sound? Probably like great audio being mangled by cheap speakers!

Antennas don't affect the way you sound, except in the sense that your signal might be pretty noisy on the receiving end if your antenna is poor. The type of antenna you choose determines how far you can talk. In other words, the better your antenna, the greater your *range*.

H-T Antennas

Thanks to repeaters, you can get away with using some truly awful antennas and still cover a decent amount of turf. The sensitivity of the repeater makes up for your lousy signal. That's why the rubber duck antennas supplied with every H-T appear to give adequate performance. In truth, they're not very good antennas, but they have the distinction of being flexible and virtually unbreakable.

Rubber ducks are fine if you expect to always be within range of a powerful repeater system. If you think you'll find yourself on the fringes, however, consider investing in one of the many telescoping $^1/_2$- or $^5/_8$-wavelength models you'll see advertised in *QST* magazine. They'll give you noticeably better coverage.

Mobile Antennas

An external antenna is the key to successful mobile operating. Yes, you can use an H-T inside a car with its rubber-duck antenna, but the results are often poor. Put a good antenna on your car and you'll likely *double* the range of your H-T! Even 50-W radios need good mobile antennas to get the best performance possible.

The antenna connects to your radio through a length of *coaxial cable* (see **Figure 2-6**). The type of coax you use for mobile applications isn't very critical because the length is so short. It *does* make a difference in home antennas, however, as you'll see later.

If you're worried about the aesthetics of your automobile, you can relax a bit.

Home Antenna Installation Tips

- Solder your coax connectors *carefully*. Too little heat will result in a poor connection, but too much heat can damage the cable. Get a veteran ham to help if you're in doubt.
- Buy an accurate VHF/UHF SWR meter and learn to use it!
- Weatherproof the coax connector after you attach it to the antenna. Cover it with a commercially available putty compound such as *Coax Seal*, or coat the connector with silicon grease and wrap the entire assembly in several tight layers of electrical tape.
- If you intend to use a chimney as an antenna support, inspect the chimney first. Make sure the bricks are still firmly in place. Even small antennas act like sails in a strong breeze. They'll add structural stress to your chimney.
- Be careful when creating a hole in your roof, siding or wherever for your coax. Make sure you're not about to drill into electrical lines or plumbing. Use generous amounts of silicon caulk to seal the hole once the cable is in place. Don't forget to add a *drip loop* at the point where the cable enters the house. A small U-shaped loop in the cable will keep rain from seeping into your home.
- When installing the coax, don't attempt sharp bends that could crimp the cable. This will change the impedance of the cable at the crimp and cause a mismatch. By the same token, take care not to crush the coax with fasteners, clamps and so on.
- If you're putting up a beam antenna, be sure to calibrate the direction the antenna is pointing with the indication on the controller in your home. (Having a friend available during this step is a big help.) When the controller says that your antenna is pointing west, for example, you want to know you can trust it!

Indoor Antennas

Antennas always work best when they're exposed to the great outdoors, but many of us are not lucky enough to have a convenient tree, chimney, balcony railing or other outside support. You might even live in a domicile where outdoor antennas are *persona non grata*! What's an eager ham to do?

There's no reason why you can't install your antenna *indoors*. Does your house or apartment have an attic? Attics are often great locations for antennas. The attic doesn't have to be large or fully finished. I managed to squeeze a 2-meter quad antenna and a rotator into an apartment attic that was only five-feet tall at the highest point. It worked great! Omnidirectional antennas work just as well in attic installations.

If you don't have an attic, consider an unused corner of a spare bedroom, or an empty space in any other room. For instance, you can hang a simple 2-meter J-pole antenna (available from several manufacturers) from a curtain rod. It won't be the best antenna you've ever used, but you'll notice a definite improvement. Nearby metal objects, including house wiring and plumbing, may *detune* indoor antennas. This simply means that you may have to fiddle with the tuning assembly on the antenna (if it has one) or adjust the length to obtain an SWR reading below 2:1.

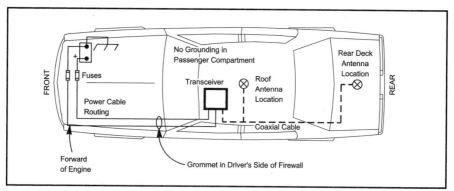

Figure 2-6—When installing a mobile antenna, route the coax to your radio by slipping it behind seats and under carpets. At the same time, keep the length as short as possible. If you're using a mobile transceiver (as opposed to a hand-held), attach the power cable directly to the car battery and put fuses in both leads.

When it comes to VHF and UHF mobile antennas, we're not talking about a 15-foot monstrosity strapped to your bumper. Most FM mobile antennas are fairly short, typically less than a few feet long. (Mobile antennas for 440 MHz are just *inches* long.) You'll find mobile antennas for single bands or multiple bands. The 2-meter/440-MHz antennas are especially popular among owners of dual-band radios.

The Longer the Better?

As you browse the mobile antenna ads you'll see references to $^1/_4$, $^1/_2$ and $^5/_8$-wavelength antennas. The greater the fraction, the longer the antenna. Most hams assume that a $^5/_8$-wavelength antenna would perform better than a $^1/_4$-wavelength antenna. This is true to a certain extent, but it depends on your operating environment.

Figure 2-7 shows the radiation patterns for typical $^1/_4$- and $^5/_8$-wavelength antennas. The patterns show the approximate directions your signals will travel as they leave these antennas. Notice how the $^1/_4$-wavelength pattern is more-or-less circular. It's radiating your signal uniformly in all directions. The $^5/_8$-wavelength antenna concentrates much of your signal energy at low angles. The low-angle pattern is great for hitting distant repeaters and other stations. However, if you have hills, buildings or mountains in the way, the low-angle energy may be blocked. With the more uniform pattern of a $^1/_4$-wavelength antenna, you stand a somewhat better chance of being heard when the terrain is rugged.

For some hams the choice is based on looks alone. They simply prefer the low profile of the $^1/_4$-wavelength design. A $^1/_4$-wavelength antenna also has a

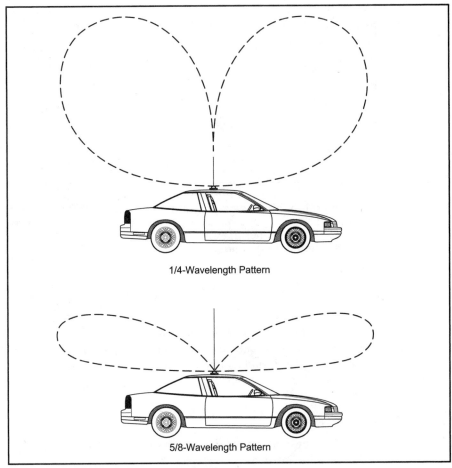

1/4-Wavelength Pattern

5/8-Wavelength Pattern

Figure 2-7— The real difference between a ¹/₄- and ⁵/₈-wavelength mobile antenna involves the radiation patterns. A ¹/₄-wavelength antenna has a more-or-less uniform pattern. A ⁵/₈-wavelength antenna radiates your signal in a pattern that sends the energy toward the horizon.

better chance of avoiding collisions with low parking-garage roofs!

Attaching Antennas to your Car

No, super glue won't do the job. You have to find a way to anchor your antenna in the 55+ MPH breezes that blow around your automobile when you're on the road. Fortunately, the manufacturers have solved this problem for you by creating a variety of mobile antenna mounts . . .

Mag mounts—A powerful magnet holds the antenna to the body of your car. You can set it up on your roof or trunk in seconds—and remove it just as quickly. The disadvantage of mag mounts is that they may scratch your car's finish. Also, you must bring the coaxial cable from the antenna to the radio through a partially open window, or by some other means. (I've heard of hams who get away with simply closing a car door on the coax, but this isn't the best approach.)

Trunk-lip mounts—These are semipermanent mounts that attach to your trunk lid, often on the hinged side. The antenna screws or snaps onto the mount. Trunk-lip mounts work fine, as long as you have enough space between the edge of the trunk lid and the body of the car when the trunk is closed. If not, the mount is probably going to scrape the body every time you open or close the trunk.

Body mounts—There's nothing complicated about these. You simply drill a large hole in the body of your precious, expensive automobile and install the antenna—permanently. (I can hear some of you shuddering at the mere *thought* of doing this.) In terms of achieving a durable installation, this is clearly the best way to go. It does tend to reduce the resale value of your car, though.

On-glass mounts—These are actually complete antenna systems, not just mounts. On-glass antennas are *très chic* among the cellular telephone set. You're simply not hip if you don't have an on-glass antenna. The ham versions look—and work—the same. The RF energy passes from one metal plate to another *through*

The classic magnetic mount mobile antenna. The base is nothing more than a powerful magnet that holds the antenna securely to the car. Note how the coax has to squeeze under the truck lid to enter the car.

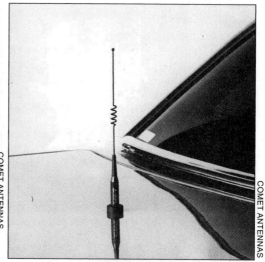

If you don't mind drilling large holes in your car, this is one of the cleanest installations you can manage.

the window glass. Although they perform reasonably well, they can be tricky devils to install. The situation is complicated further if you own a car with a window defroster that uses a thin metallic membrane in the glass. RF energy doesn't pass through this stuff very well!

Home Antennas

Do you want to radiate your home signal in all directions at once? If your answer is "yes," you want an omnidirectional antenna. These antennas are popular on VHF and UHF because they're easy to install and require little, if any, adjustment. If you install an omni antenna as high as possible, you'll enjoy plenty of contacts.

Popular omnidirectional base designs include the *groundplane*, *discone* and *J-pole*. Like mobile antennas they are available in single and multiband versions. Most VHF/UHF omni designs are relatively short, but others tower to heights of 15 to 20 feet. It all depends on the frequency (the lower the frequency, the longer the antenna) and the design.

There are times when you may need to focus your signal energy, particularly when you need to reach a distant station. You can't do that very well with an

(Right) This multiband omnidirectional antenna is known as a *discone*.

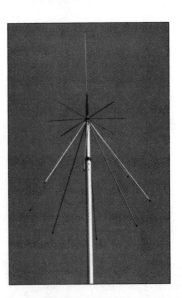

(Left) An omnidirectional antenna such as this one will give you decent coverage in all directions. Note how the antenna (a *collinear ground plane*) attaches to a short mast, which is secured to the chimney with stainless-steel straps.

omnidirectional antenna. Instead, you need a *beam* design. The most popular type of beam antenna for VHF and UHF is the *Yagi* (see **Figure 2-8**). The runner-up is the *quad* (see **Figure 2-9**).

Both of these antennas concentrate your signal (transmit *or* receive) in specific directions. They accomplish this by bouncing the energy back and forth between various *elements*. In a Yagi, the elements are pieces of metal attached to a long boom. Quad elements are wires cut to square shapes and supported away from the boom. The more elements a beam antenna has, the more it concentrates your signal (and the longer it is, too).

Signal concentration in a beam antenna can make up for the lack of power in a transceiver. Let's pretend you're using an H-T with only a 2-W output. A big beam can focus those two measly watts and make it sound as though you're running 200 W!

If beams are so great, why doesn't everybody have one? Well, the focusing ability of a beam is a double-edged sword. A beam antenna that's pointed north, for example, is a poor antenna for signals coming from any other direction. If you want to talk to someone south of your station, you must turn the antenna so it points south. Now we're talking about yet another device and another purchase: An *antenna rotator*.

Rotators are nothing more than fancy electric motors. They're designed to

Figure 2-8—A small 2-meter Yagi antenna like this one will focus your signal in the direction of your choice. Note that the antenna elements are perpendicular to the ground. This is called *vertical polarization*. Most FM operators use vertical polarization.

Figure 2-9—This impressive quad antenna will concentrate your power just like a Yagi.

turn heavy loads clockwise or counterclockwise. If cable TV hasn't taken over your neighborhood, you may still see rooftop television antennas equipped with rotators. The rotators are controlled by a selector box that's usually somewhere close to the TV. A three or five-wire cable connects the rotator to the selector box. When you twist the selector knob, the rotator dutifully obeys and turns the antenna to the requested direction.

TV antenna rotators are light-duty devices, but they're usually powerful enough to turn a moderate-sized VHF or UHF beam. For larger beam antennas, you may need to invest in a heavy-duty rotator designed for ham applications. Ham rotators may also come with fancy control electronics complete with digital readouts and so on.

TRANSMISSION LINES—THE CRITICAL LINKS

You own a great radio. You've installed a terrific antenna. Now all you have to do is get the RF energy from the radio to the antenna, and vice versa. As straightforward as it sounds, this is the point where many hams go astray.

The *transmission line* is the cable that connects your radio to your antenna. Perhaps you're already familiar with it by another name: *coaxial cable*, or simply *coax*. It's called coax because of the way the cable is constructed. "Coaxial" means "two axis," like two concentric circles. The inner axis is the *center conductor*. This can be a single wire or a tight bundle of stranded wire. The outer axis is the *shield*. The shield can be made of braided wire, or solid, flexible metal. It's separated from the inner axis by plastic or some other insulating material.

"Transmission line" can also refer to cables that are not "coaxial" at all. The flat *ribbon cable* you see on some TV antennas is a type of transmission line. Transmission lines such as these are popular for hamming on the HF bands, but they're not used much for amateur VHF and UHF.

I could write an entire chapter on the subject of transmission lines, but you'd want to hang yourself from the nearest tree by the time you finished reading it. To tell the truth, transmission line theory is lethally boring. Let's cut to the chase, shall we?

In **Table 2-1** I've created a list of coaxial cable recommendations depending on your particular antenna installation. You'll notice that the types of cable I suggest change according to how much is used and the frequency in question. Even under the best conditions, transmission lines lose a little of the RF energy you put into them. This loss increases as the cable gets longer, the frequency gets higher, or both.

The suggestions shown in the table are based on the assumption that you've installed everything perfectly. The last thing you want to do is increase the losses that could occur in your coax, but there are ample opportunities to do just that!

You see, transmission lines, antennas and radios all have their own *impedance*. In simplest terms, impedance is resistance to the flow of an ac signal. RF energy is an ac signal.

If the impedances of the radio, transmission line and the antenna are the same,

Table 2-1

Recommended Coaxial Cables

Find your band of interest on the left-hand side of the table, then move right until you reach the length range you need.

Band	Length (feet)				
	<10	10-50	50-100	100-200	>200
50 MHz	RG-58	RG-8/U	8214	9913	9913
146 MHz	RG-58	8214	9913	Hardline	Hardline
222 MHz	RG-58	8214	9913	Hardline	Hardline
440 MHz	8214	9913	9913	Hardline	Hardline
1296 MHz	8214	9913	Hardline	Hardline	Hardline

you're on easy street. In the case of VHF and UHF ham equipment, that magic impedance number is usually 50 Ω (ohms). But if the impedance somewhere in the system is anything other than 50 Ω, you have a condition called a *mismatch*. A mismatch isn't necessarily a crisis. It all depends on how bad it is.

Any number of things can cause a mismatch: Poorly soldered coax connectors, a mistuned antenna, damaged cable, etc. What we care about most is the effect the mismatch has on the energy in the coax. A mismatch causes a portion of the energy to *reflect* from the location of the mismatch and go bouncing back and forth along the coax. This creates a *standing wave* condition that you can measure with an SWR (Standing Wave Ratio) meter.

When there is no mismatch at all, the SWR will be 1:1. A slight mismatch might show itself with a reading of 1.2:1, which isn't a cause for concern. If you get a reading of 2:1 or higher, however, you need to find out why. Generally speaking, the higher the SWR, the greater the losses in the cable because more of your valuable energy is wasted as it ricochets through the coax. At VHF and UHF frequencies, the loss caused by a mismatch condition can be *huge*! Your radio may also be damaged by a high SWR condition. (Many transceivers are designed to reduce their output if the SWR climbs above 2:1.)

The bottom line: Calculate the length of coax required to connect your radio to your antenna. Figure out the highest band of frequencies where you intend to operate, now or in the future. (Always plan for your *future* needs!) Go to Table 2-1 and select the suggested cable.

Of course, my cable recommendations assume that you'll install your antenna system properly. (You *will*, won't you?) This means attaching the connectors without damaging the coax, setting up the antenna and testing it according to manufacturer's instructions and so on. (See the sidebar, "Antenna Installation Tips.") The result should be an antenna system with an SWR of 2:1 or less.

TURNING UP THE POWER

If your radio is only capable of putting out a couple of watts or so, the day will probably arrive when you realize that you need more RF muscle. Mobile operating at the edge of a repeater's coverage area often makes this need painfully apparent. When you're trying to say something important (to you, at least), it's frustrating to know that your signal is weak into the repeater. You can bet that the folks who are listening to your transmissions are equally miffed!

If you can't rely on your antenna to add some punch to your signal, you'll need to work on the *power* side of the equation. You can compensate for antenna deficiencies by boosting your output to 30 W, 150 W or more. All you need is an amplifier, affectionately known as a "brick."

Zack Jackowski, KC2FNB, prefers his FM hamming from an igloo!

A Brick is a Brick is a Brick . . .

Bricks are solid-state RF power amplifiers. They're called bricks because they approximate the shape and weight of the venerable baked-mud masonry. Bricks are not terribly complicated. They're basically brute-force amplifiers designed to take your flea power and convert it to tiger power—or at least to a power level equivalent to that generated by a small burrowing mammal.

The average brick features an ON/OFF switch that allows you to bypass the amplifier when you're feeling overwhelmingly overconfident about your signal. Some include a receive preamplifier.

On some bricks you'll find a switch labeled SSB/FM. Don't let it confuse you. This ominous selector only affects the amplifier's *hang time*.

When you're operating SSB, your output power changes with the amplitude of your voice. If you're not talking, there's no output. Bricks go into action the instant they sense RF power from your radio. As you babble, the RF power is coming and going at a rapid clip. (Every time you pause to draw a breath, for example, your output drops to nearly nothing.) This causes the hapless brick to jump from transmit to receive like a toad in a hail storm. Relays and other components don't react well to this sort of abuse.

The solution is to make the amplifier remain in the transmit mode for a second or so *after* the output falls to zero. When you flip the switch to SSB, the amp waits briefly before returning to the receive mode—just to make sure you don't have more to say.

On FM, your radio is supplying full output for as long as you hold down the microphone push-to-talk button. The amp stays solidly in the transmit mode until you decide to listen once again. So, you don't want the brick to hesitate before it switches to receive. Leave it in the FM position and the amp will bounce back to receive the instant you release the mike button.

How Much is Enough?

How much output power do you really need? Thirty to 50 W is usually sufficient for mobile work. There are plenty of power amps on the market that will take 5 W or less and kick it up to 30 W. Brand new units in this power class will cost between $60 and $160, depending on how many features you want. Of course, you could run 150 W or more from your car, but I wouldn't recommend it. The intense RF is likely to drive your automotive electronics crazy while it's rearranging your DNA.

For hearth and home the question of power depends on what you want to do. If your goal is to put a solid signal into all the local repeaters with your omni antenna, you can probably remain at the 30-W level. But let's say you enjoy operating FM simplex and you want to expand your coverage. Maybe you've just been elected as the control station for the Klingon Language Net. This is no time to soft-peddle your signal. A jump to the 150-W class may be in order.

YOU'RE ON THE AIR!

To use a repeater, you must know one exists. There are various ways to find these elusive beasts. Local hams can provide information about repeater activity, or you can consult a repeater listing. The ARRL publishes two guides that are indispensable to the active repeater user: *The ARRL Repeater Directory* and a nifty software package called *Travel Plus*. They're available from your favorite dealer, or directly from ARRL Headquarters.

Most modern FM transceivers also include scanning features that will let you sweep up and down the band to your heart's content. This is another easy way to find active repeaters. Listen in the early morning and late afternoon.

Remember that you're listening to repeater *output* frequencies. The frequencies you transmit on to use the repeaters—the *input* frequencies—will be different. Many FM radios sold within the last several years handle the differences between input and output frequencies automatically. For example, if you hear a repeater on 147.36 MHz, your radio will automatically select 147.96 MHz as its transmit frequency. The rig "knows" that most repeaters that operate above 147 MHz in the 2-meter band have their input frequencies higher than their outputs. It also knows that the 2-meter input/output frequency separation is 600 kHz. (147.96 MHz −147.36 MHz = .60 [600 kHz])

Naturally, there are exceptions to every rule. You might find a repeater on the 2-meter band that's using something other than a 600-kHz offset. Look up the repeater in *The ARRL Repeater Directory* or check *Travel Plus* and you'll discover its input frequency. Program the input/output frequency pairs into an available memory channel and you're in business.

Don't Grab the Microphone Yet

Finding a repeater is only half the job. Spend some time really *listening* to it. How busy is it? What sort of conversations do you hear? Do the people sound friendly?

Most importantly, learn the *customs* of the repeater. When someone is fishing for a contact, how do they announce themselves? Is it, "Hey! You stupid fools! I wanna talk to someone!" or something more subtle such as, "WB8IMY, monitoring"? (You'll find that most repeater operators use the words "monitoring" or "listening" as a way to let everyone know that they're in the mood to chat.)

Bruce Steele, KCØDZB, takes his 2-meter FM hand-held transceiver to high locations such as Scotts Bluff National Monument in Nebraska.

Where's the FM Action?

When you're looking for repeaters, set your radio to scan between the following frequencies . . .

6 Meters:
   51.62—51.98 MHz
   52.5—52.98 MHz
   53.5—53.98 MHz

2 Meters:
   145.20—145.50 MHz
   146.61—147.39 MHz

222 MHz:
   223.85—224.98 MHz

440 MHz:
   442.00—445.00 MHz
   447.00—450.00 MHz

1296 MHz:
   1282—1288 MHz

For FM simplex activity, try these national calling frequencies:
   52.525 MHz, 146.52 MHz,
   223.5 MHz, 446 MHz, 1294.5 MHz

Now, With Trembling Hands . . .

Push must eventually come to shove. The immovable object must meet the unstoppable force. In other words, you must *talk* to someone! You punch in the frequency of the local repeater and listen. Silence. This is the moment of truth. You key the microphone and, in your most confident voice, announce your call sign.

The repeater transmits for a few seconds, then stops. Surely someone is reaching for their microphone. They'll call you in just a few seconds . . . won't they? The seconds stretch into minutes. You announce that you're listening again, this time with added urgency.

Still nothing.

Again the lonely minutes pass. Maybe you just picked a bad time. You'll try again in an hour or so. As you reach for the POWER switch, the repeater suddenly comes to life.

"WB8ISZ this is WB8SVN. You around, Dave?"

"WB8SVN from WB8ISZ. I'm here. Did you just get off work?"

Now you feel a new emotion—anger! It's a safe bet that one of these two guys were listening before. Why didn't they answer you? Is it because you're a new ham?

The Shy Communicators

Hams pride themselves on their ability to communicate, yet there is an odd contradiction: many hams are painfully shy! If you don't believe this, go to any hamfest. Chances are, you'll see hams whose call signs you recognize—hams who are constantly chattering on the local repeaters. So why are these same hams wandering around so quietly? When you approach them, why do they seem so ill at ease and reluctant to talk?

The answer lies in the nature of Amateur Radio itself. With the exception of visual modes such as ATV, no one can see you when you're on the air. You could be holding a conversation with someone while wearing little more than your underwear. They'd never know! In other words, ham radio allows us to hold the world at arm's length while still maintaining contact. It acts as a filter and a shield for those who are uncomfortable with close, personal communications.

Breaking through the shyness barrier to communicate with a stranger is difficult. Think back to your school days. When the teacher asked for student volunteers for a project, why did you hesitate? Perhaps you wanted to see if anyone else was willing to join you. No one wants to be the first to raise their hand!

A similar situation occurs on repeaters. When you announced that you were listening, a dozen people may have heard you. No one recognized your call sign, though. You're a stranger, an unknown. It's as though the teacher just got on the repeater and asked for volunteers to speak to you. Who will be the first to step forward?

For many hams, the familiar line of reasoning is, "Hmmm . . . I don't know this guy. What would I say to him? Nah . . . I'll wait. I'm sure someone else will give him a call." The problem is, when all the hams on the repeater feel this way, no one replies!

And so it goes on repeaters throughout the country. The problem isn't you *per se*, it's that fact that you're a stranger. So how do you make the transition from stranger to friend?

Breaking the Ice

If you keep announcing that you're "listening," someone is bound to come back to you eventually. This could take a long time—especially if you're trying to start a conversation during less popular hours. To really break the ice and shed your "stranger" label, you need to assert yourself on the air. That is, you need to become part of an existing conversation.

Listen to the repeater during the early morning and late afternoon. That's when it's likely to be used the most. As you hear stations talking to each other,

Dave Rosenthal, N6TST, operates from a cruise ship off the coast of Baja California.

listen for an opportunity to contribute something—even if it's just a question. Let's say that you find two hams discussing computers . . .

"KR1S from WR1B. Well, I'm definitely going to pick up some extra memory at the show tomorrow. I figure I need at least 64 megabytes."

"I don't know, Larry. I think 128 megs would be a better choice for the kind of software you're running."

Even if you don't own a computer, I bet you can think of a question that will give you an excuse to join the conversation. In the pauses between their transmissions, announce your call sign.

"WB8IMY"

"Well, there's a new voice. Ah . . . WB8IMY . . . I think it was . . . this is KR1S. How can I help you?"

"Hello. My name is Steve and I live in Wallingford. I'm thinking about buying a computer for my Amateur Radio station, but I'm a little confused. You guys seem knowledgeable. Can you give me a recommendation?"

Perfect! Stroking a person's ego is the best way to get them talking. With luck, these fellows will be more than happy to show off their expertise. Just keep the questions and comments coming.

If you engage in enough of these conversations on the same repeater, you'll gradually melt through the shyness barrier. In time, your call sign will be as familiar as any other. When you say, "WB8IMY listening," you'll have a much better chance of getting a response. After all, they'll *know* you.

Getting Involved

Another way to establish yourself is to become involved in club activities. Look for a local club that's active in public-service events. Attend the meetings regularly and be prepared to volunteer whenever they ask for help.

Don't worry about your lack of experience in public-service operating. Believe me, it isn't that difficult. You'll be told exactly what to do and, in most cases, an experienced ham will be nearby.

My first public-service activity was a canoe race in my home town of Dayton, Ohio. I was the new face in the club and I was new to ham radio. When they asked for volunteers, it took a great deal of courage to raise my hand. Boy, am I glad I did!

The race organizers needed "checkers" at various points along the river. It was our task to make sure that each canoe passed our checkpoint safely. I was stationed with my FM transceiver at an isolated rural bridge over the Miami River. As each canoe passed beneath me, I checked it off my list and relayed the information to the net-control station. The sun was shining, a gentle breeze was blowing through the trees and I felt terrific! Here I was, an Amateur Radio operator, doing an important job with my fellow team members.

After the race, we all met at a local pizza restaurant and swapped stories. Someone asked if I wanted to be part of the communications team for the March of Dimes walk-a-thon the following weekend. Why not? After participating in several public-service events, everyone knew me by name and call. There was never a shortage of someone to talk to on the repeater.

Acknowledging Stations

If you're in the midst of a conversation and a station transmits its call sign between transmissions, the next station in queue should acknowledge that station and permit the newcomer to make a call, or join the conversation. It's discourteous not to acknowledge him and it's impolite to acknowledge him but not let him speak. You never know; the calling station may need to use the repeater immediately. He may have an emergency on his hands, so let him make a transmission promptly.

The Pause That Refreshes

A brief pause before you begin each transmission allows other stations to participate in the conversation. Don't key your microphone as soon as someone else releases his. If your exchanges are too quick, you'll block other stations from getting in.

The Silent Service

Subaudible means "below audible"; below the range of human hearing. Repeaters use these low-frequency audio tones for special purposes. Your ears can't hear them, but a repeater has no problem detecting their presence. Subaudible tone generators and decoders are often lumped together under the term "continuous tone-coded squelch system," or CTCSS for short. Modern FM transceivers include CTCSS generators (encoders), or at least provide them as options.

Most repeaters use subaudible tones as an effective way to deal with interference. When a repeater is using a CTCSS detection system, it will only repeat signals that carry the proper subaudible tone. An interfering signal, such as *intermod* caused by a nearby commercial transmitter, will be ignored. So, if you can't seem to use a particular repeater—and you're sure you are within its range—it might be using a CTCSS system. You'll need to find the correct subaudible tone and program it into your transceiver. Some radios will scan for CTCSS tones on a repeater's signal and display the tone frequency. If your radio lacks this neat feature, check for tone information in *The ARRL Repeater Directory*, or ask around at your next club meeting.

The W1KKF repeater in Wallingford, Connecticut, uses subaudible tones in an interesting way. The problem centers on a powerful repeater in the New York City area that shares the same frequencies with W1KKF. Although the users of the New York repeater don't usually key up W1KKF, their repeater is powerful enough to be heard throughout a large part of W1KKF's coverage area when the band is open. As you can guess, anyone with a sufficiently sensitive receiver is driven insane by signals from two repeaters at once!

The solution? W1KKF transmits a 162.2-Hz subaudible tone whenever it repeats a signal. Hams who own radios with CTCSS squelches can set their rigs to respond only to signals that carry the 162.2-Hz tone. In other words, they won't hear a peep unless the signal is coming from the W1KKF repeater.

The "courtesy beepers" on some repeaters compel users to leave spaces between transmissions. The beep sounds a second or two after each transmission to permit new stations to transmit their call signs in the intervening time period. The conversation may continue *only after the beep sounds*. If a station is too quick and begins transmitting before the beep, the repeater may respond to the violation by shutting down! So, if the repeater uses a courtesy beep, wait until you hear it before you continue talking.

Brevity is the Soul of Wit

Keep each transmission as short as possible. Short transmissions permit more people to use the repeater. All repeaters promote this practice by having timers that "time-out," temporarily shutting down the repeater if someone babbles beyond the preset time limit. With this in the back of their minds, most users keep their transmissions brief!

Learn the length of the repeater's timer and stay well within its limits. The length may vary with each repeater; some are as short as 15 seconds and others are as long as three minutes. Some repeaters automatically vary their timer length depending on the amount of activity on the system; the more activity, the shorter the timer.

Because of the nature of FM radio, if more than one signal is on the same frequency at one time, it creates a muffled buzz or an unnerving squawk. If two hams try to talk on a repeater at once, the resulting noise is known as a "double." If you're in a roundtable conversation, it is easy to lose track of which station is next in line to talk. There's one simple solution to eradicate this problem forever: *Always pass off to another ham by name or call sign.* Saying, "What do you think, Jennifer?" or "Go ahead, 'YUA" eliminates confusion and avoids doubling. Try to hand off to whoever is next in the queue, although picking out anyone in the roundtable is better than just tossing the repeater up for grabs and inviting chaos.

Autopatching

Autopatches allow mobile and portable operators to place telephone calls through the repeater. Let's say I'm cruising around town and I see a car accident. By pressing three keys in sequence on my H-T's keypad, I can activate the repeater autopatch. I'll hear the dial tone when I release the push-to-talk (PTT) switch. Now I can transmit again, this time using the keypad to dial 911.

Although they may seem similar, a repeater autopatch is not the same as a cellular telephone. They both use RF, but the similarity ends there. Autopatches are comparable to old-fashioned "party lines." When you make an autopatch call, *everyone* gets an earful of your conversation. Cell phones are relatively private by comparison. Autopatches are only *half-duplex* devices. This means that you and the person you've called must take turns talking. While you're babbling away, your buddy can't get a word in whatsoever. Cell phones, on the other hand, are *full duplex;* you can interrupt each other at will. Finally, autopatches are intended for short-term use. You make your call, speak your piece and get off. You can chat all day on a cell phone—if you can recover from the shock when you see your bill.

Just because you're able to use a repeater, don't expect free access to its autopatch. Most repeater groups require autopatch users to be paid members. As a member in good standing, you'll receive the "secret" codes that operate the autopatch. Let's say that you've just paid your dues and you've been told that the repeater autopatches codes are *-9-1 to turn it on and #-3-6 to turn it off. Here's how an autopatch conversation might sound . . .

"WB8IMY to access the 'patch"

(I announce my intention and listen for a second, just to make sure that no one needs to use the repeater for an emergency.)

*-9-1

(While holding down the microphone button, I press the access code on the radio keypad. I release the mike button and I hear the repeater sending a telephone dial tone.)

5-5-5-6-9-3-8

(I hold down the mike button again and punch in the telephone number. As I release the mike button, I hear the telephone ringing.)

"Hello?" my friend answers.

"Hi, Alan. This is Steve. I'm calling you by ham radio from my car. We have to take turns talking, okay? I can't hear you when I'm speaking."

"I understand."

"I'm about 15 minutes from your house. Mind if I stop by and pick up that old mobile antenna?"

"That's fine. I'll have it ready. Can you stay for lunch?"

"I'd be glad to. See you soon."

"Bye, Steve"

#-3-6 (I send the code that shuts down the autopatch.)

"WB8IMY clearing the 'patch." (Just another courtesy to let everyone know that I'm finished.)

Free Speech?

Can you use an autopatch for *any* purpose? Well, consider that you're actually using the facilities of someone else's Amateur Radio station. That's exactly what a repeater is, remember? The control operator has the last word about what takes place on his or her repeater—including the autopatch.

This is another instance where it pays to learn the customs of your repeater. For example, some control operators don't mind if you use the autopatch to order a pizza. Others mind very much! Listen for a while and learn the pattern. If in doubt, don't do it until you can speak to the control operator.

Beyond the preferences of the control operator is the mighty arm of the FCC. When it comes to autopatches, the FCC has set up several legal taboos.

1. Thou shalt not use an autopatch in the commission of a crime. ("Hey,

Charlie! I'm just a couple of blocks away. Can I pick up that kilo of coke you promised me?")

2. Thou shalt not use an autopatch to avoid long-distance telephone charges. ("It's long distance, but don't worry. I'm using a ham autopatch.")

3. Thou shalt not use an autopatch for business purposes. ("This is Gwen. Has our shipment of banana leaves arrived on the dock yet? Don't forget to check the bill of lading before you accept it.")

Simplex Patches

Simplex autopatches use a single frequency to provide a link between mobile or portable transceivers and local telephone systems. Unlike repeater autopatches that operate in full duplex applications (transmitting and receiving simultaneously on separate frequencies), a simplex autopatch uses the same frequency for both transmission *and* reception.

Many simplex autopatches accomplish this feat by cycling from transmit to receive in fractions of a second. During each brief receive cycle, the patch checks the radio for the presence of a carrier. If a carrier is detected, it assumes that the operator of the mobile or portable station is talking and it feeds this audio to the telephone line. If not, it continues to transmit the telephone audio while monitoring the receiver for activity. (Simplex autopatches include timers that will shut down the system if nothing is heard from the receiver after a certain amount of time.)

Assuming that the transceiver can cope with such rapid switching, it's possible to carry on a conversation with minimal interference—although each party is likely to hear pulsing "clicks" in the audio. Both parties must also take turns talking. Like a repeater autopatch, a simplex autopatch is controlled through the use of DTMF (*TouchTone*) tones sent from your transceiver keypad.

Simplex autopatches are often used in situations where hams want telephone access without the cost and hassle of building a complete repeater system. A simplex autopatch requires only a transceiver, an antenna and the means to connect to the telephone line. They don't provide the smooth, reliable operation of a repeater autopatch, but they're better than having no autopatch facilities at all.

With declining prices, simplex autopatches are growing in popularity. The problem, however, is finding available frequencies for them. Operating a simplex autopatch amounts to remote control of a station, so you must use them at 222.15 MHz or above. Setting up an autopatch system on whatever frequency strikes your fancy is a formula for trouble.

Reverse Autopatches

A *reverse autopatch* is exactly what the name implies. It accepts incoming calls from the telephone system and then sends a signal over the air (through the repeater) to indicate that someone has called. By responding with the proper tones from their radios, any hams who are monitoring the repeater can access the autopatch and answer the call.

The user of a reverse autopatch is essentially controlling the repeater remotely (via the telephone lines). If the caller is a ham, this doesn't present a problem—assuming that all other legal requirements have been met. But if the caller is *not* a ham, you suddenly have a situation where a nonamateur is in control of a ham station without a control operator present. The FCC expressly forbids this!

Some repeaters get around this problem by using a kind of voice mail system. A nonham can call the repeater and leave a message. No signal is sent over the air. It's up to the hams to periodically access the repeater and check their voice mail to see if any messages are waiting.

3 | A World of Bits, Bytes...and Radios

If you had a file that you wanted to transfer to a friend's computer several miles away, how would you do it? I bet you'd make a copy on a diskette or CD and deliver (or send) it to your friend, right? Fair enough.

But what if you were too lazy to drive over to your friend's house, or too cheap to spring for a few postage stamps? (I don't mean to imply that you have a dysfunctional personality. This is just an analogy!) You could hook up a *modem* to your computer and, assuming your buddy has a modem as well, send the file to him over the telephone line.

Modems simply take data from your computer—in the form of shifting voltages—and transform it into shifting audio tones. Once you've made this magical transmutation, it's easy to send the information over the telephone. After all, if a telephone line can send your voice, it can send just about any other type of audio signal. Depending on the speed your modem is running, the tones sound like a chorus of buzz saws or hissing tomcats. The telephone lines handle these raucous tones just fine.

The same modems also receive tones from telephone lines and translate them back to shifting voltages. When fed to your computer, these shifting voltages are interpreted as data. There you have it! Two computers with modems can communicate just about anything between each other . . . as long as a telephone line is available.

Snip! Snip!

Oops! There goes that telephone line! Too bad. How will you get the data to your buddy now?

Well, you're a ham and he's a ham. You have those fancy transceivers sitting on your desks. Can you use a radio link to replace the telephone line? You bet!

RADIO MODEMS

You can substitute Amateur Radio for your telephone lines—all you need is some type of radio modem. Radio modems come in several flavors and names depending on the application. TNCs or terminal *node controllers* are designed primarily for use with VHF and UHF FM transceivers. Their specialty is a communication mode known as *packet* where your data is sent in short "packets" of information. More about packet in a moment.

Another type of radio modem is the *multimode communication processor*, or MCP. As the name implies, this is a versatile device that's capable of running in several different digital modes—most of which we'll discuss in this chapter. See **Figure 3-1**.

Finally, your own computer can function as a radio modem if it's equipped with a sound card. The sound card translates data into audio tones, and audio tones into data. See **Figure 3-2**. With the proper software, your computer can also become a "multimode" radio modem.

With so many radio modems at your disposal, let's find out what you can do with them!

AUTOMATIC POSITION REPORTING SYSTEM (APRS)

Imagine staring at a map on your computer monitor. It's a map of your state. Blue lines indicate rivers and other bodies of water. Green lines trace major highways. You press a key and zoom into a particular area of your state. Up and down the river you see clusters of symbols with Amateur Radio call signs. As you watch,

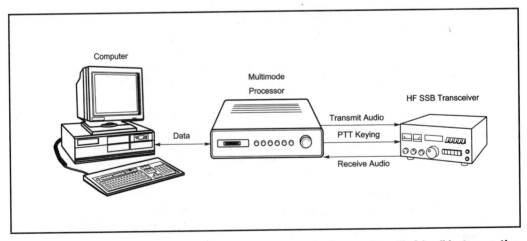

Figure 3-1—A multimode communication processor is the modem "bridge" between the computer and the radio.

Multimode Communication Processors

Kantronics, 1202 East 23rd St, Lawrence, KS 66046; tel 785-842-7745;
www.kantronics.com

MFJ Enterprises, PO Box 494, Mississippi State, MS 39762; tel 601-323-5869;
www.mfjenterprises.com/

HAL Communications, 1201 W. Kenyon Rd, Urbana, IL 61801-0365; tel 217-367-7373;
www.halcomm.com/

Timewave Technology Inc, 58 Plato Blvd E., St Paul, MN 55107; 651-222-4858;
www.timewave.com/

SCS, Roentgenstr 36, D-63454 Hanau, Germany; **www.scs-ptc.com/**

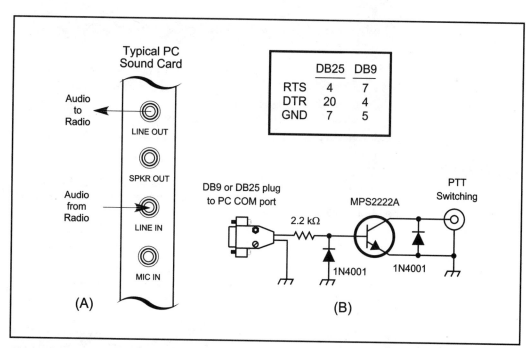

Figure 3-2—You can also use your computer sound card as a radio modem by connecting audio lines to and from your radio (A). A simple transistor switching circuit (B) allows your computer's COM port to switch your radio from transmit to receive.

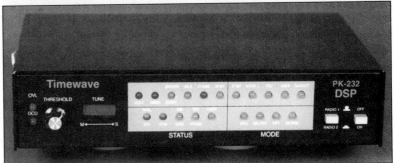

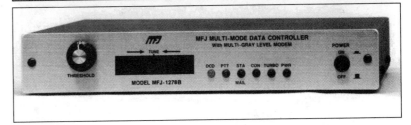

Popular multimode communications processors include (top to bottom), the Kantronics KAM Plus, the Timewave PK-232/DSP and the MFJ-1278B.

your packet TNC receives a transmission . . . and one of the symbols *moves!*

To quote Dorothy from *The Wizard of Oz*, "Toto, I don't think we're in Kansas anymore!"

But if Dorothy had been carrying an Automatic Position Reporting System (APRS) in her basket, we'd at least be able to see where she was! We would even be able to follow the path she takes on her way to the Emerald City. All we'd have to do is follow the Yellow Brick . . . er, the computer-generated map.

APRS was the brainchild of Bob Bruninga, WB4APR. APRS exploits the ability of a TNC to transmit *beacon packets* that carry short strings of alphabetical characters or numbers. A beacon is an *unconnected* packet. You can think of unconnected packets as "broadcasts." The information is sent to no one in particular and can be received by anyone. Other stations can relay an unconnected packet as well.

The ability to send beacons has been a feature of TNCs since the earliest days of packet when hams used them to send messages back and forth to each other over Amateur Radio "packet networks." Internet e-mail made the ham network obsolete, but two developments in the '90s finally made APRS the new "king" of VHF/UHF data communication. The first was the introduction of small TNCs that could operate on battery power. These little boxes could go anywhere without relying on ac power. The second was the debut of compact, affordable *Global Positioning System* (GPS) receivers. GPS receivers rely on signals from military satellites to determine your position to an accuracy of ±100 feet. Most of these receivers have ports that allow their data output to be transferred to other devices . . . such as TNCs. See **Figure 3-3**.

It didn't take long for Bob to realize that he was on to something big. If you

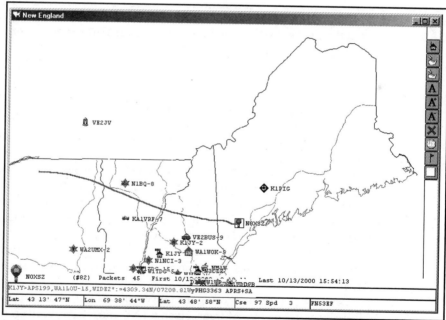

In the autumn of 2000, manned balloons lifted off from New Mexico for a cross-country race. Each balloon carried an Amateur Radio APRS system with a GPS receiver. I was able to track the NØXSZ balloon as it floated west to east from New York into southern Maine (see the sloping dark line in this *WinAPRS* display). Note the other icons that represent APRS stations in homes and in cars.

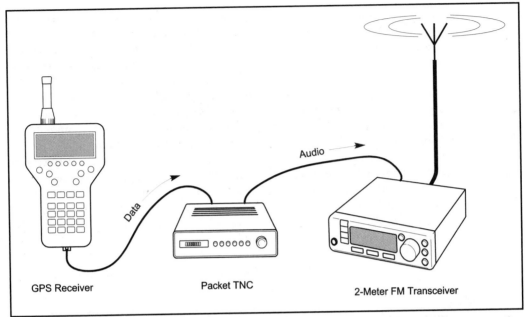

Figure 3-3—A typical APRS system with a GPS receiver. The precise wiring between the GPS receiver and the packet TNC can vary, so check your TNC's manual for details.

could take the position information from a GPS receiver and incorporate it into beacon packets transmitted by a TNC, you could tell everyone on the network *exactly where that GPS receiver was located.* And if the TNC could pass that information along to a computer that could display the position by using a symbol on a map . . . bingo! The Automatic Position Reporting System was born.

Although APRS' mapping capability was developed to display the movement of hand-held GPS receivers, most features evolved from earlier efforts to support real-time packet communications at special events. Any person in the network, upon determining where an object is located, can move his cursor and mark the object on his map screen. This action is then transmitted to all screens in the network, so everyone gains, at a glance, the combined knowledge of all network participants!

Let's say you're monitoring the movements of rafts during a river race. Each rafter carries a 2-meter FM transceiver, a TNC and a GPS receiver. If your station picks up a transmission from any raft along the river, it will automatically relay the information to everyone else. So, everyone's maps are continually updated with the latest positions of the rafts.

You don't need to buy a GPS receiver to enjoy APRS. All you need is the

APRS software (see the sidebar, "Software on the Web") and a packet TNC to function as your radio modem. Just determine your latitude and longitude as best you can. Look it up in an atlas, or borrow a friend's GPS receiver just long enough to determine the position of your station. After you feed your position information to the software, your TNC will regularly announce your position to anyone else who is monitoring. You can even use APRS to exchange bulletins and enjoy live conversations with others on the network.

You should be aware that GPS receivers have been plunging in price. Starting out at well over $1000, you can now pick up GPS receivers for less than $200. The trick is that you must use a GPS receiver that features a "NEMA compatible" data output port. As I've already mentioned, this is the jack that feeds the data to your packet TNC. When you place it in the APRS mode, the TNC will look for the data stream from the receiver and extract the data it needs. Then, the TNC will transmit that position data over your FM radio. (Most APRS activity is on 2 meters, with 144.39 MHz being the primary frequency.) Just make sure the TNC you buy for APRS includes APRS *firmware*. The firmware contains the instructions for operating in the APRS mode. The good news is that most modern TNCs include APRS firmware.

Software on the Web

BlasterTeletype (RTTY)
**www.geocities.com/SiliconValley/
    Heights/4477/**

DigiPan (PSK31)
members.home.com/hteller/digipan/

DSP-CW: (CW and RTTY)
www.zicom.se/dsp/index.html

IZ8BLY Hellschreiber (Hellschrieber)
iz8bly.sysonline.it/. Also, **members.
    xoom.com/ZL1BPU/software.html**

Mix32W (RTTY and PSK31)
tav.kiev.ua/~nick/my_ham_soft.htm

Multimode (RTTY and PSK31 for
    Macintosh computers. PowerPCs
    recommended.)
    **www.blackcatsystems.com/software/
    multimode.php3**

PSK31 (for Linux)
aintel.bi.ehu.es/psk31.html

RITTY
Brian Beezley, K6STI, 3532 Linda Vista
    Dr, San Marcos, CA 92069;
    k6sti@n2.net. $100 with delivery via
    e-mail, $5 additional for postal delivery.
    Check or money order only.

Stream (MFSK16 and other modes)
iz8bly.sysonline.it/

TrueTTY (RTTY)
www.dxsoft.com/

WinPSK (PSK31)
www.winpskse.com/

WinAPRS (APRS for *Windows*)
aprs.rutgers.edu

APRS on the Air

If you are interested in APRS, my suggestion is that you spend some time just watching the activity in your area. Connect a shielded audio cable between the external speaker jack of your VHF FM radio (even a scanning receiver will do) and the audio input of your chosen TNC. You'll also need to connect a serial cable between your TNC and your computer.

Set your radio on 144.39 MHz and boot up your APRS software. (Actually, you may want to take time beforehand to read the APRS software manual.) "Open" your VHF TNC in the APRS software. In other words, open the lines of communication between your TNC and the software. There is a specific menu item in the software to do this. Finally, zoom the APRS map display on your area of the country.

As the beacon packets arrive at your radio, your software should respond by displaying location "icons" on the map. (Be patient. This could take a few minutes.) You can click on these icons to learn a little more about the stations that are sending the beacons. Some of these stations even have weather instruments and you can observe their recorded temperatures, wind speeds and more!

In time your computer-generated map should begin to fill with icons. Some will be fixed stations, but others will be mobiles and you can watch their positions change as they move. If you zoom your map out, you may notice that you are seeing icons from stations thousands of miles away. How is this possible on VHF?

Remember that every APRS station is capable of relaying beacon packets. The very distant packets probably found their way into Internet *wormholes* (data routes between APRS stations that use the Internet) and were relayed to a station in your region, which in turn repeated the packets to the stations in your vicinity

Anson, N9RJX, took this APRS setup—consisting of a GPS receiver (right) and a 2-meter hand-held transceiver (left) on a friend's business jet while flying to the Dayton Hamvention.

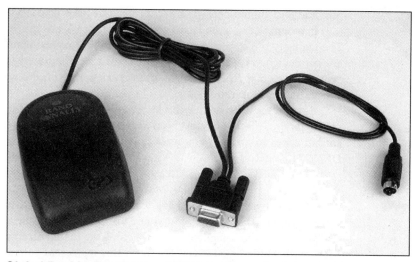

Global Positioning System (GPS) receivers don't have to be expensive. Rand McNally sells this little unit as part of the *StreetFinder* software package for less than $100. Connect the GPS receiver to an Amateur Radio packet TNC and you're ready for go-anywhere APRS.

and, ultimately, to you.

To learn more about APRS, pick up a copy of *APRS: Tracks, Maps and Mobiles* by Stan Horzepa, WA1LOU at your favorite dealer, or from the ARRL. You can download APRS software on line. See the sidebar "Software on the Web."

RTTY

Radioteletype, better known as *RTTY*, is the granddaddy of digital communication below 30 MHz (the HF bands) and remains the mode of choice for digital contesting and DXing.

Unlike packet radio, RTTY does not use any form of error detection; what you see on the screen is what you get. The goal isn't to transfer information error free, as we discussed at the start of this chapter. Instead, with RTTY we simply want to get enough information across to be understood—human to human. Even so, with adequate demodulator sensitivity and sharp filtering, it's possible to enjoy excellent copy under poor conditions.

The traditional road to RTTY has been through multimode communication processors such as those manufactured by Kantronics, MFJ, Timewave, HAL Communications and others. With a computer and an SSB radio, these devices make it possible to communicate digitally throughout the world.

In recent years, sound card software for RTTY has made substantial inroads. The majority of these programs are intended for "casual operating." That is, they are not designed for competitive RTTY such as DX pileups or contesting. A few programs, such as *RITTY 4.0* by Brian Beezley, K6STI, *are* written to meet exacting performance requirements.

Beyond the software or the MCP, all you need for RTTY (and all other HF digital modes, for that matter) is an SSB transceiver. High power and big antennas are not necessary unless you want to nail down first place in a contest, or on the DXCC Honor Roll.

RTTY is casual and fun. After you've hooked up all the necessary hardware and software, it is a matter of finding someone. Most RTTY activity takes place on the 20-meter band between 14.080 and 14.095 MHz. Listen for the distinctive *blee-blee-blee* rythmns of the RTTY signal. Put your SSB transceiver in the lower sideband mode (LSB) and tune the signal carefully, watching the tuning display on your software or MCP. With a little practice you'll see text flowing across your monitor.

Watch the conversation. You'll notice how one station transmits, then the other. They say "over to you" (it's your turn) using Morse code abbreviations such as BTU K ("back to you, over"). A short RTTY exchange might go like this:

WB8IMY DE N6TQH . . . YES, I DID BUY THAT NEW RADIO TODAY, BUT I HAVE NOT HAD A CHANCE TO SET IT UP YET. DIDN'T YOU BUY A NEW RIG AS WELL? WB8IMY DE N6TQH BTU K

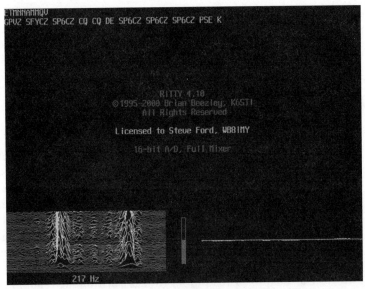

RITTY is an good example of high performance RTTY software for *DOS*.

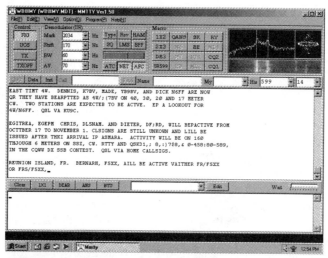

MMTTY is RTTY software for *Windows*.

N6TQH DE WB8IMY . . . GOOD NEWS! I AM EAGER TO HEAR YOUR NEW TRANSCEIVER ON THE AIR. NO, I DID NOT PUCHASE A NEW TOY MYSELF. I HAVE TO SAVE MY MONEY. N6TQH DE WB8IMY K

Consult your software or MCP manual on how to set up your SSB transceiver to transmit RTTY. The most common method is to use the accessory jack that you'll find on the rear panel of almost any modern SSB radio. At this jack there are connections for receive audio, transmit audio (from your MCP or sound card) and PTT (push to talk) keying. The PTT connection is used by your computer or MCP to put your radio into the transmit mode when it is your turn to send text.

Chasing RTTY DX

You'll have plenty of opportunities to chat with hams in other countries on RTTY. You can work 100 countries to earn your RTTY DX Century Club (DXCC) award, and even climb the ladder to the stratospheric heights of RTTY DXCC Honor Roll. RTTY DXing is competitive and highly rewarding.

Like any other form of DXing, the quest for RTTY DX demands patience and skill. When a DXpedition is on the air with RTTY from a rare DXCC entity, your signal will be in competition with thousands of other HF digital operators who want to work the station as badly as you! Sometimes pure luck is the winning factor, but there are several tricks of the trade that you can use to tweak the odds in your favor.

Don't Call...Yet

Let's say that you're tuning through the HF digital subbands one day and you stumble across a screaming mass of RTTY signals. On your computer screen you see that everyone seems to be frantically calling a DX station. Oh, boy! It's a pileup!

You can't actually hear the DX station that has everyone so excited, but what the heck, you'll activate your transceiver and throw your call sign into the fray, right? *Wrong!*

Never transmit even a microwatt of RF until you can copy the DX station. Tossing your call sign in blindly is pointless and will only add to the pandemonium. Instead, take a deep breath and wait. When the calls subside, can you see text from the DX station on your screen? If not, the station is probably too weak for you to work (don't even bother), or he may be working "split." More about that in a moment.

If you can copy the DX station, watch the exchange carefully. Is he calling for certain stations only? In other words, is he sending instructions such as "North America only"? Calling in direct violation of the DX station's instructions is a good way to get yourself blacklisted in his log. (No QSL card for you—ever!) Does he just want signal reports, or is he in the mood for brief chats? Most DX stations simply want "599" and possibly your location—period. Don't give them more than they are asking for. (A DX RTTY station on a rare island doesn't care what kind weather you are experiencing at the moment.)

Working the Split

When DX RTTY pileups threaten to spin out of control, many DX operators will resort to working "split." In this case, "split" means split frequency. The DX station will transmit on one frequency while listening for calls on another frequency (or range of frequencies).

A good DX operator will announce the fact that he is working split with almost every exchange. That's why it is so important to listen to a pileup before you throw yourself into the middle. If you tune into a pileup and cannot hear the DX station, tune below the pileup and see if you copy the DX station there. If his signal is strong enough, he shouldn't be hard to find if he is working split. His signal will seem to be by itself, answering calls that you cannot hear. This is a major clue that a split operation is taking place. Watch for copy such as...

CQ DX DE FOØAAA, UP 10 (He is listening up 10 kHz)

CQ DX DE FOØAAA, 14085-14090 (He is listening between 14085 and 14090 kHz)

Whatever you do, *never call a split DX operation on the station's transmitting frequency.* Your screen will quickly fill with rude comments from others who are listening. Instead, put your transceiver into the split-frequency mode (better drag out your rig manual!). Set your radio to receive on the DX station's transmitting frequency and transmit on his listening frequency. If the DX station is listening through a range of frequencies, you'll need to select a spot where you think you'll be heard. Change your transmit frequency if this particular "fishing spot" doesn't seem to be working.

Short and Sweet

When you've done your listening homework and you're ready for battle, by all means fire at will. Wait until the DX station finishes an exchange. He'll signal that he is ready for another call by sending "QRZ?" or something similar. When it's time to transmit, make it short and to the point, like this…

WB8IMY WB8IMY WB8IMY K

Listen again. Has he responded to anyone yet? If the answer is "no" and other stations are still calling, the DX is probably trying to sort out the alphabet soup of confusion on his screen. Fire again!

WB8IMY WB8IMY WB8IMY KK

You might make it through at just the right moment when other signals subside briefly, or when the ionosphere gives you an unexpected boost. But if no one is calling, or if you hear the DX station calling someone else, *stop*. You lost this round, so give the lucky winner his chance to be heard. Your next opportunity will be coming up shortly.

Did He Call You?

Watch your screen carefully. If the DX station only copied a fragment of your call sign, he might send something like…

IMY IMY AGAIN??

In this instance he copied only the last three letters of my call sign ("IMY"). Fading and flutter can make RTTY signals difficult to copy clearly. The best thing to do is reply right away, sending your call sign three times just like before.

If luck is on your side, you'll see…

WB8IMY DE FO0AAA … TNX. 599. QSL? KK

And with excited fingers you answer…

QSL. UR 599. TNX AND 73 DE WB8IMY K

PSK31

PSK31 has several characteristics in common with RTTY. It is not an error-free digital mode, but is designed for casual keyboard-to-keyboard conversation. Best of all, PSK31 works its magic with only about 30 Hz of spectrum. This is a major advantage on the crowded HF digital subbands.

Peter Martinez, G3PLX, invented PSK31. For a few years, PSK31 languished in obscurity because special DSP hardware was necessary to use it. But in 1999, Peter designed a version of PSK31 that needed nothing more than a common computer sound card. It was a simple piece of software that ran under *Windows* and used the sound card as its interface to the transceiver. Peter made the software available at no cost on the Internet. Announce that you are offering free software to the ham community and the reaction will be predictable—PSK31 took off like gangbusters.

The PSK31 community received another jolt in 2000 with the debut of "panoramic" software such as *DigiPan* and *WinPSK*. Both software packages made it outrageously simple to get on the air with PSK31—all you have to do is hook up the necessary cables, then point and click your mouse.

In a span of just two years, PSK31 has become the *Number One* HF digital mode for casual keyboard-to-keyboard operating. It has also been embraced enthusiastically by the QRP (low power) community—and for good reason. With just a couple of watts and a wire antenna you can work stations throughout the United States, along with a good selection of DX as well. PSK31 is easy to operate and the software is inexpensive (free, in most cases). With most amateurs owning sound-card-equipped computers these days, that's a combination too powerful to resist.

PSK31 Software

The first step in setting up a PSK31 station is to jump onto the Web and download the software you need, according to the type of computer system you are using. As this book went to press there were PSK31 programs available for *Windows* (98, 95 and 3.1), *Linux*, *DOS* and the Mac. The sidebar "Software on the Web" lists Web addresses for software downloading.

Once you have the software safely tucked away on your hard drive, install it and read the "Help" files. Every PSK31 program has different features—too many to cover in this chapter.

The Panoramic Approach

All of the PSK31 programs are good; there isn't a bad apple in the bunch. But when I was introduced to *DigiPan*, it was love at first sight.

One of the early bugaboos with PSK31 had to do with tuning. Most PSK31 programs required you to tune your radio carefully, preferably in 1-Hz increments. In the case of the original G3PLX software, for example, the narrow PSK31 signal would appear as a white trace on a thin *waterfall* display. Your goal was to

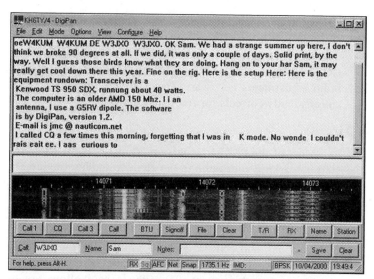

bring the white trace directly into the center of the display, then tweak a bit more until the phase indicator in the circle above the waterfall was more-or-less vertical. Regardless of the software, PSK31 tuning required practice. You had to learn to recognize the sight and sound of your target signal. With the weak warbling of PSK31, that wasn't always easy to do. And if your radio didn't tune in 1-Hz increments, the receiving task became even more difficult.

Nick Fedoseev, UT2UZ and Skip Teller, KH6TY, designed a solution and called it *DigiPan*. The "pan" in *DigiPan* stands for "panoramic"—a complete departure from the way most PSK31 programs work. With *DigiPan* the idea is to eliminate tedious tuning by detecting and displaying not just one signal, but entire *groups* of signals.

If you are operating your transceiver in SSB without using narrow IF or audio-frequency filtering, the bandwidth of the receive audio that you're dumping to your sound card is about 2000 to 3000 Hz. With a bandwidth of only about 31 Hz, a lot of PSK31 signals can squeeze into that spectrum. *DigiPan* acts like an audio spectrum analyzer, sweeping through the received audio from 100 to 3000 Hz and showing you the results in a large waterfall display that continuously scrolls from top to bottom. What you see on your monitor are vertical lines of various colors that indicate every signal that *DigiPan* can detect. Bright yellow lines represent strong signals while blue lines indicate weaker signals.

The beauty of *DigiPan* is that you do not have to tune your radio to monitor any of the signals you see in the waterfall. You simply move your mouse cursor to the signal of your choice and click. A black diamond appears on the trace and *DigiPan* begins displaying text. You can hop from one signal to another in less than a second merely by clicking your mouse! If you discover someone calling CQ and you want to answer, click on the transmit button and away you go—no radio adjustments necessary. (And like the original PSK31 software, *DigiPan* automatically corrects for frequency drift.)

On the Air

Most of the PSK31 signals on 20 meters are clustered around 14.070 MHz, start by parking your radio in the vicinity and booting up *DigiPan*. **Do not touch your rig's VFO again.** Just place your mouse cursor on one of the vertical signal lines and right click. That's all there is to it!

PSK31 signals have a distinctive sound unlike any digital mode you've heard on the ham bands. You won't find PSK31 by listening for the *deedle-deedle* of a RTTY signal, and PSK31 doesn't "chirp" like the TOR modes. PSK31 signals *warble*—that's the best way I can describe them.

One remarkable aspect of *DigiPan* is that it allows you to see (and often copy) PSK31 signals that you cannot otherwise hear. It is not at all uncommon to see several strong signals (the audible ones) interspersed with wispy blue ghosts of very weak "silent" signals. I've clicked on a few of these ghosts and have been rewarded with text (not error free, but good enough to understand what is being discussed).

Using *DigiPan* reminds me of the sonar operators in the movie *The Hunt for Red October*. There is an eerie excitement in finding one of those ghostly traces and muttering to yourself, "Hmmm...what do we have here? An enemy submarine rigged for silent running? A distant pod of killer whales? Or Charlie in Sacramento running 5 W to his attic dipole?"

Spend some time tracking down PSK31 signals and watching the conversations. With a little practice you'll discover that tuning becomes much easier. You'll also find that you develop an "ear" for the distinctive PSK31 signal.

A PSK31 Conversation

Conversing with PSK31 is identical to RTTY. For example:

Yes, John, I'm seeing perfect text on my screen, but I can barely hear your signal. PSK31 is amazing! KF6I DE WB8IMY K

I know what you mean, Steve. You are also weak on my end, but 100% copy. WB8IMY DE KF6I K

Some PSK31 programs and processor software offer type-ahead buffers, which allow you to compose your response "off line" while you are reading the incoming text from the other station. The original PSK31 software for *Windows* lacked this feature, although it did allow you to send "canned" pre-typed text blocks known as *brag files*. (Brag files are usually descriptions of your station set up.)

PSK31 conversations flow casually, just like RTTY. The primary difference is that you will usually experience perfect or near-perfect copy under conditions that would probably render RTTY useless.

MFSK16

MFSK is really a type of super-RTTY. The MFSK technique was developed during the heyday of teleprinter HF communications, as a way to combat multi-path propagation problems, providing reliable point-to-point communications with relatively simple equipment.

What does this new mode consist of? Well, there are 16 tones, sent one at a time at 15.625 baud, and they are spaced only 15.625 Hz apart. Each tone represents four binary bits of data. The transmission is 316 Hz wide, and has a CCIR specification of 316HF1B. It is exactly like RTTY, but with 16 closely spaced tones instead of two wider spaced tones. With a bandwidth of 316 Hz, the signal easily fits through a narrow CW filter.

The signal has an amusing musical sound, is quite narrow, clean to tune across, and not unpleasant to listen to. The sound is certainly better and the bandwidth narrower than some modes on HF these days!

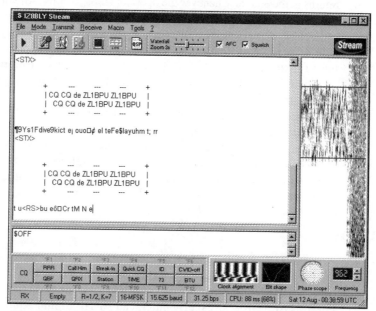

Stream, by Nino Porcino, IZ8BLY, was the first *Windows* software for MFSK.

To get started with MFSK16 you'll need a sound-card equipped computer (in addition to the HF SSB radio). Next you'll need the IZ8BLY *Stream* program (see the sidebar, "Software on the Web"). Downloading the *Stream* software and installing it is very simple. Fortunately the help information is also available as a separate download, so you can read that before you install. *Stream* offers a generous collection of tools along the top, separate transmit and receive windows, a good collection of definable "macro" buttons, and an excellent "waterfall" tuning display. Along the bottom is a list of settings and parameters, plus the date and time. There is also a drop-down log window, for automatic logging and insertion of QSO information, and a very useful "QSP" window for relaying incoming text.

The software is very simple to use—start typing and it transmits, and press **F12** on your computer keyboard to end the transmission. The trouble comes in tuning in the MFSK signals. It takes some skill and a certain patience learning to tune in MFSK, but the results are worth the effort.

Because the tones are closely spaced and the filters very narrow, you must have a very stable transceiver, and you must use the tuning provided with the software, not the transceiver tuning. The software allows you to tune up and down in 1 Hz steps, or click on the waterfall for exact tuning. The waterfall has a zoom function, and zoom ×3 is best.

Tuning is done using the waterfall display. Under the lower horizontal line (red on the screen) you'll see a broad band towards the left. This is the idle carrier, the lowest of the 16 tones. This carrier is transmitted briefly at the start of each over, and returns at the end, or whenever the operator stops to think. All you need to do is center the red line on this carrier, and the AFC will keep it there. During the over, you'll see little black vertical stripes all over the waterfall, with gray "sidelobes" above and below. These are the transmitted symbols, and once again, you can adjust the software tuning so the red line centers the lowest of these symbols. Unfortunately while this is easy when the signal is already tuned, finding the correct spot on a weak signal during an over is not so simple and takes a little practice.

Once you've found the right spot, almost perfect text will start to appear on the screen, although it is delayed by some 3 to 4 seconds as the data trickles through the error correction system, and appears one or two words at a time.

The mode is a delight to use once you learn to tune in. The typing speed is fast, and while changeover from transmit to receive is not as fast as RTTY, it is quite good enough for excellent casual conversations.

PACTOR

PACTOR strolled onto the telecommunications stage in 1991. It combined the best aspects of packet (the ability to pass binary data, for example) and added robust *error-free* capability (one of the few HF digital modes that can make this boast.) PACTOR was eagerly embraced by HF digital equipment manufacturers and became the most popular HF digital communication mode in a remarkably

WinLink 2000—Internet E-mail from Anywhere!

The Internet has become the e-mail medium of choice for most hams, but there is a sizeable group of amateurs who often travel beyond the reach of the Internet. This group includes hams at sea, travelers in recreational vehicles (RVs), missionaries, scientists and explorers. No doubt the day will come when wireless, affordable Internet e-mail access will be available from any point on the globe. Until that day arrives, however, Amateur Radio HF digital operators have a very capable substitute!

More than 21 HF digital stations worldwide have formed a remarkably efficient e-mail network known as WinLink 2000. Running *WinLink 2000* software and using primarily PACTOR or PACTOR II, these facilities transfer e-mail between HF stations and the Internet. They also "mirror" (share) messages between themselves using the Internet, allowing amateurs to pick up their e-mail from any WinLink 2000 station. Thanks to the WinLink 2000 network, HF digital operators at sea, on the road or anywhere else can exchange Internet e-mail with nonham friends and family.

WinLink stations scan a variety of HF digital frequencies on a regular basis, listening on each frequency for about two seconds. By scanning through frequencies on several bands, the WinLink stations can be accessed on whichever band is appropriate according to your location and the propagation conditions at the time.

You can access Winlink 2000 stations using just a basic PACTOR setup. However, most users also rely on a piece of software known as *Airmail* to handle uploading and downloading automatically. *Airmail* is a 32-bit program that runs under *Windows-95, 98* or *NT 4.0*. Airmail supports the SCS PTC-II and IIe PACTOR-2 processors, as well as the Kantronics KAM+ and KAM-98, AEA/Timewave PK-232 and PK-900 modems, and the MFJ 1276 and 1278B. You can download a copy of *Airmail* online at **www.airmail2000.com**. To learn more about WinLink 2000, see K4CJX's Web site at **www.winlink.org/k4cjx/**.

---

short period of time. PACTOR was also widely adopted for mailbox operations and other forms of message handling. Today it still remains the most popular of the *error-free* modes used below 30 MHz.

Most PACTOR is done using standalone multimode processors like the MFJ, Kantronics, HAL or Timewave products I've already mentioned.

PACTOR II debuted in the mid '90s. Like its little brother PACTOR, PACTOR II uses DSP techniques and innovative data coding to achieve extraordinary error-free performance. PACTOR II is only available in multimode processors manufactured or licensed by Special Communications Systems (SCS), and they tend to be expensive ($800). This has slowed PACTOR II's acceptance in the ham community. In 1999, SCS introduced a pared-down processor (the PTC-IIe) that offered the same performance, but at a somewhat lower cost ($650).

To operate PACTOR (either flavor) you'll need an SSB transceiver that can switch from transmit to receive and back again *very* quickly (see the sidebar "The Need for Speed"). Although PACTOR operators take turns transmitting during their conversations (like RTTY, PSK31 and MFSK16), they maintain continuous

The Need for Speed

If you are considering PACTOR, it is critical that your HF transceiver be able to switch from transmit to receive very quickly.

Why, you ask?

All of the burst modes use some form of ARQ—automatic repeat request. In the basic system, a chunk of data is sent and then the sending station waits for *a specific amount of time* to hear from the receiving station. Was everything received without errors? If the answer is "yes," the receiving station transmits an acknowledgment signal, or ACK, and the next data chunk is sent. If the answer is "no," a non-acknowledgment, or NAK, is transmitted and the data is repeated. This sets up a kind of ARQ dance where the stations ping-pong back and forth until everything makes it through error free. For the dance to work properly, however, the transmitting station must hear the ACKs and NAKs. If the rig at the transmitting station does not switch fast enough, the ACK or NAK could arrive before the it is ready to receive. We're talking *milliseconds* of time!

The rule of thumb is to look for a radio that can switch from transmit to receive in less than 30 ms. The lower the number, the better. *QST* Product Reviews often measure transmit/receive-switching times for exactly this reason.

links with each other by sending short bits of information back and forth—even when they are not typing on their keyboards. This creates a *chirp-chirp-chirp* signal that reminds some hams of electronic crickets. The station computers are "speaking" to each other. They are saying, in essence, "Did you receive the last transmission from me?" The other station replies, "No, I didn't. Send the data again," or "Yes, I got it. Send the next chunk of data." That's what is going on when you hear all that chirping on the air!

PACTOR isn't typically used for conversations, contesting or DXing, although you can find the occasional "live" PACTOR operator. PACTOR is chiefly used these days to communicate with automated stations to transfer messages and other information (see the sidebar, "WinLink 2000—Internet E-mail from Anywhere!").

4 | The Mystery of the "Weak Signals"

Look beyond hamming on the VHF bands with just your FM mobile or hand-held transceiver. What do you see? Is the rest of the VHF and UHF amateur spectrum a trackless desert . . . or a fertile land rich in exciting opportunities?

The answer depends on your expectations. If you expect a nonstop gabfest whenever you switch on the radio, you'll be gravely disappointed. If you expect consistently strong signals and clear communications, you'd better turn to another chapter.

But . . .

If you're the type of person who enjoys taking the road less traveled, the type of person who revels in the unknown, this may be one of the most important chapters in the book!

"WEAK SIGNALS"?

If you listen to veteran hams talking about single sideband (SSB) and CW activity on the VHF/UHF bands, you might hear it referred to as "weak signal" operating. Doesn't sound very encouraging, does it? I mean, who wants to have a weak signal? Where's the excitement in weak signals?

The term "weak signal" is a bit misleading. It means that you're often dealing with signals that have traveled great distances, losing much of their energy along the way. Compared to the sledgehammer signals you get from your local repeater, these are weak indeed. In most cases you need beam antennas to concentrate the feeble energy and perhaps a preamplifier to boost the sensitivity of your receiver. SSB and CW are used because these modes are most efficient when you're communicating directly on VHF and UHF. SSB and CW signals are detectable at levels where FM signals can't even be heard.

Out of the Woodwork

One of the secrets to enjoying SSB and CW operating on VHF and UHF is knowing how to search for contacts. Unless there is a contest going on, you're unlikely to find a conversation the moment you flip the POWER switch. There's too much space and too few operators. To make the contacts, you have to work *smart.*

The first step is knowing how everyone else is operating, and to follow their lead. Essentially, this means to listen first. Pay attention to the segments of the band already in use, and follow the operating practices the experienced operators are using.

How are the Bands Organized?

In most areas of the country, everyone uses *calling frequencies* to establish contact. Then the two stations move up or down the band to chat. This way, everyone can share the calling frequency

W4RXR's VHF portable site in the Great Smoky Mountains.

without having to listen to each other's conversations. You can easily tell if the band is open by monitoring the call signs of the stations making contact on the calling frequency. A complete list of calling frequencies is shown in **Table 4-1**.

SSB/CW activity on VHF/UHF is concentrated on the two lower VHF bands, 6 and 2 meters (50 and 144 MHz). The number of active stations on these bands is about equal. Above the 2-meter band, there are considerably more active stations on the 70-cm (432 MHz) band than any other.

On 6 meters, a *DX window* has been established to reduce interference to DX stations. Yes, *DX* stations! During years of high solar activity, 6-meter openings to the other side of the world are possible! Even during the "quiet" years 6 meters will occasionally open for contacts spanning 2000 miles or more.

The window, which extends from 50.100-50.125 MHz, is intended for DX contacts only. The DX calling frequency is 50.110 MHz. US and Canadian 6-meter operators should use the domestic calling frequency of 50.125 MHz for nonDX work. When contact is established, move off the calling frequency as quickly as possible.

Activity Nights

Although you can scare up a contact on 50 or 144 MHz almost any evening

Table 4-1		
VHF/UHF Calling Frequencies		
50.125 MHz		
144.20 MHz		
222.10 MHz		
432.10 MHz		

Table 4-2		
Common Activity Nights		
Band (MHz)	*Day*	*Local Time*
50	Sunday	6:00 PM
144	Monday	7:00 PM
222	Tuesday	8:00 PM
432	Wednesday	9:00 PM
902	Friday	9:00 PM
1296 and up	Thursday	10:00 PM

(especially during the summer), in some areas of the country there isn't always enough activity to make it easy to find someone. Therefore, informal *activity nights* have been established. There's a lot of variation in activity nights from place to place. Check with an active VHFer near you to find out about local activity nights. See **Table 4-2**.

Contacts *can* be made on non-activity nights as well. It may just take longer to get someone's attention.

Local VHF/UHF *nets* often meet during activity nights. (A *net* is a group of hams who meet on the air to exchange information or discuss various topics.) Two national organizations, SMIRK (Six Meter International Radio Klub) and SWOT (Sidewinders on Two), run nets in many parts of the country. These nets provide a meeting place for active users of the 50 and 144-MHz bands. For information on the meeting times and frequencies of the nets run by SMIRK or SWOT, ask other occupants of the bands in your area, or see the Info Guide for more information.

What's that Beeping?

If you tune around on 6 meters when propagation conditions are good, you'll probably hear several *beacon* stations. Beacons send their call signs and other information in slow-speed Morse code. Other information may include their *grid-square* locations (see below), output power and antenna height. Most beacons use nondirectional antennas and relatively low power. If you can hear a beacon in a certain geographic area, you can probably work stations in that area. If you hear a beacon signal from several hundred or several thousand miles away, the band is open. Point your beam toward the beacon and start calling CQ! (Don't call CQ on the beacon frequency, though.) There are beacons on the bands above 50 MHz also, and they operate in a similar fashion. An extensive list of VHF beacons appears in *The ARRL Operating Manual*.

Grid Squares

One of the first things you'll notice when you tune the low end of any VHF band is that most conversations include an exchange of *grid squares*. Grid squares are a shorthand means of describing your general location anywhere on Earth.

(For example, instead of trying to tell distant stations that, "I'm in Wallingford, Connecticut," I tell them, "I'm in grid square FN31." It sounds strange, but FN31 is a lot easier to locate on a map.)

Grid squares are coded with a 2-letter/2-number/2-letter code (such as FN24kp). This handy designator uniquely identifies the grid square and your exact location in latitude and longitude; no two have the same identifying code.

There are several ways to find out your own grid square identifier. The first bit of information you need is the *approximate* latitude and longitude of your station. (In most cases, the latitude and longitude of your city or town will be sufficient.) Your town engineer can provide this information, or you can go to a library and check a couple of geographic atlases. A nearby airport is another good source.

If you're fortunate enough to own a *GPS* (global positioning system) receiver, you can get precise latitude and longitude information in seconds. If you don't have one of these wonders lying around, perhaps you know a friend who does.

With data in hand, you're ready to determine your grid square. If you have a copy of *The ARRL Operating Manual,* see Chapter 12. It includes a grid square map similar to the one shown in **Figure 4-1**. It also tells you how to convert your latitude and longitude to a grid square designator.

Propagation

If you're new to the world above 50 MHz, you might wonder what sort of range is considered "normal." To a large extent, your range on VHF is determined by your location and the quality of your station. For example, a high-power station with a stack of beam antennas on a 100-foot tower will outperform a 10-W rig and a small antenna on the roof.

But for the sake of discussion, consider a more-or-less "typical" station. On 2-meter SSB, a hypothetical typical rig would be low-powered, perhaps a multimode transceiver (SSB/CW/FM), followed by a 100-W amplifier. The antenna of our typical station might be a single 15-element Yagi at around 50 feet, fed with low-loss coax.

Using SSB or CW, how much territory could this station cover on an average evening? Location plays a big role, but it's probably safe to say you could talk to similarly equipped stations about 200 miles away almost 100% of the time. Naturally, higher-powered stations with high antennas have a greater range, up to a practical maximum of about 350-400 miles in the Midwest (less in the hilly West and East).

On 222 MHz, a similar station might expect to cover about the same distance, and somewhat less (perhaps 150 miles) on 432 MHz. This assumes normal propagation conditions and a reasonably unobstructed horizon. This range is a lot greater than you would get for noise-free communication on FM. Increase the height of the antenna to 80 feet and the range might extend to 250 miles, and probably more, depending on your location. That's not bad for reliable communication!

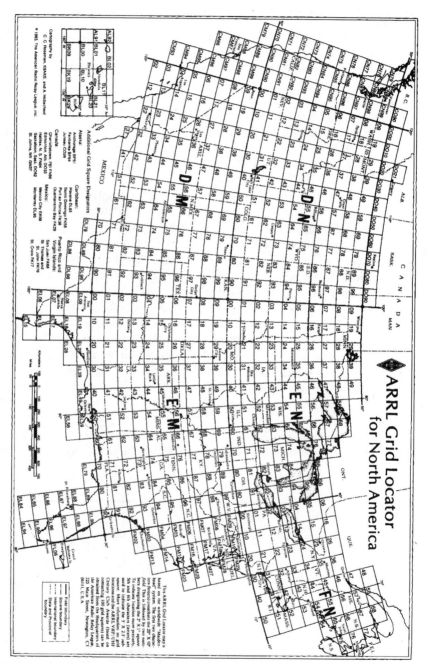

Figure 4-1—A grid-square map of the United States, similar to the one that appears in *The ARRL Operating Manual*. Grid-square designators are exchanged to qualify for contest points and various awards.

Band Openings and DX

The main thrill of the VHF and UHF bands is the occasional band opening, when signals from far away are received as if they're next door! DX of well over 1000 miles on 6 meters is commonplace during the summer, and occurs at least a few times each year on 144, 222 and 432 MHz.

DX propagation on the VHF/UHF bands is strongly influenced by the seasons. Summer and fall are definitely the most active times although band openings occur at other times as well. Here is a review of the most popular types of VHF/UHF propagation. Remember that there is a lot of variation, and that no two band openings are alike. This uncertainty is part of what makes VHF/UHF interesting and fun!

• *Tropospheric—or simply "tropo"—openings.* Tropo is the most common form of DX-producing propagation on the bands above 144 MHz. It comes in several forms, depending on local and regional weather patterns. This is because it is caused by the weather. Tropo may cover only a few hundred miles, or it may include huge areas of the country at once. The best times of year for tropo propagation are from spring to fall, although they can occur anytime. One indicator of a possible tropo opening is dew on the grass in the evening. Another is a high-pressure weather system stalled over or near your location.

• *Meteor scatter* communication uses the ionized trails meteors leave as they pass through the atmosphere. VHF radio signals can be reflected by these high-altitude meteor trails and return to Earth hundreds or even thousands of miles away (see **Figure 4-2**). This ionization lasts only a second or so. Most meteor-scatter contacts are made on 6 and 2 meters. Because the ionization from a single meteor is brief, special operating techniques are used. (See the sidebar, "Hooked on Meteors.")

Meteor-scatter contacts are possible at any time of year. Activity is greatest during the major meteor showers, especially the Perseids, which occurs in August.

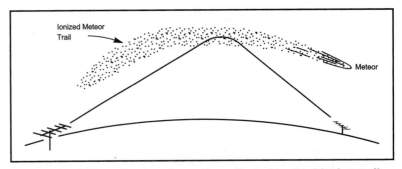

Figure 4-2—VHF radio signals can be reflected by the blazing trails of meteors as they enter our atmosphere. The results are quick contacts with stations hundreds or thousands of miles away.

• *Sporadic E* (abbreviated E_s) propagation is the most spectacular DX producer on the 50-MHz band, where it may occur almost every day during late June, July and early August (see the sidebar "How Good Can It Get?"). A short E_s season also occurs during December and January. Sporadic E is more common in mid-morning and again around sunset during the summer months, but it can occur at any time and any date. E_s also occurs at least once or twice a year on 2 meters in most areas. E_s results from small patches of ionization in the ionosphere's E layer. E_s signals are usually strong, but they may fade away without warning.

• *Aurora* (abbreviated Au) openings occur when the auroras are sufficiently ionized to reflect radio signals. Auroras are caused by the Earth intercepting a massive number of charged particles from the Sun. Earth's magnetic field funnels these particles into the polar regions. The charged particles often interact with the upper atmosphere enough to make the air glow. Then we can see a visual aurora. The particles also provide an irregular, moving curtain of ionization that can propagate signals for many hundreds of miles.

Aurora-reflected signals have an unmistakable ghostly sound. CW signals sound hissy; SSB signals sound like a harsh whisper. FM signals refracted by an aurora are often unreadable. (Score another one for SSB and CW!)

• *EME, or Earth-Moon-Earth* (often called *Moonbounce*) is the ultimate VHF/UHF DX medium. Moonbouncers use the Moon as a reflector for their signals, and the contact distance is limited only by the diameter of the Earth (both stations must have line of sight to the Moon). See **Figure 4-3**. As you've probably guessed, Moonbouncers have a particular obsession about knowing where the Moon is, especially when they can't see it because of cloud cover. Barking at the Moon indeed!

Moonbounce conversations between the USA and Europe or Japan are commonplace—at frequencies from 50 to 10,368 MHz. That's true DX! Hundreds of

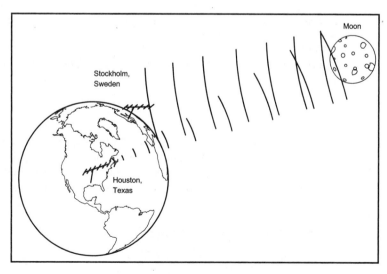

Figure 4-3—By reflecting signals off the surface of the Moon, a ham in Houston, Texas, can communicate with a ham in Stockholm, Sweden. This technique is known as *moonbounce*, or EME. If one station is using large antennas and high power, the other station can be very modest (100 W and a single beam antenna).

Hooked On Meteors!

By Tom Hammond, WD8BKM

With three meteor contacts under my belt, I think I'm hooked!

I'd been active on 2-meter SSB for about three years, but I had never tried to make a meteor-scatter contact. The idea of bouncing my signal off the flaming tail of a meteor seemed absurd. Only the big-gun stations did that kind of stuff. My modest 2-meter setup (a Yaesu FT-736R, a small amplifier and a single Yagi) just didn't have enough oomph. Or did it?

Setting Up

It started Monday night, December 12, 1994, while I was listening to the Cincinnati-based VHF/UHF Net. I was hearing meteor bursts with astonishing frequency. The signals would pop out of nowhere for several seconds, then vanish utterly. If I could hear them, could I also work them? I jumped into the net and arranged a meteor schedule with AD4FF for the upcoming Geminids shower. We'd try to make contact Wednesday night on 144.143 MHz.

The next day, I posted electronic messages to the Internet Usenet groups rec.radio.amateur.misc and rec.radio.amateur.space requesting meteor skeds. I don't have HF equipment, so scheduling contacts on 75 meters (as many scatter buffs do) was not an option. I was happy to get a response from Rupert, N2OTO, on Long Island. We exchanged e-mail and set up a sked for 0330 UTC on December 14, following the sked with AD4FF.

Tuesday night, I listened on 144.200 MHz (the national SSB calling frequency) from 0230 UTC onward. The frequency was buzzing (or should I say "bursting"?) with activity. The meteor bursts were loud, but very short. I heard only partial call signs. I quickly decided that random meteor skeds should be left to the pros.

Timing is Everything

Timing is critical in meteor work. The typical procedure is to divide each minute into 15-second intervals. One station transmits for the first and third interval. The other station transmits for the second and fourth interval.

Each station calls the other giving both call signs (for example, "AD4FF WD8BKM"), repeating the calls over and over during the transmit interval. When you hear a burst with both call signs, you respond with a report indicating the length of the burn in seconds, such as "S2." The other station responds with, "Roger S2." The contact is complete when you exchange "rogers." I like to finish with, "Roger 73," if time permits.

Meteor scatter doesn't support much in the way of conversation, but it's a fascinating

---

EME-capable stations are now active, some with gigantic antenna arrays. Their antenna systems make it possible for stations running 100 W and one or two Yagi antennas to work them. Activity is constantly increasing. In fact, the ARRL sponsors an EME contest, in which Moonbouncers compete on an international scale.

Hilltopping and Portable Operation

Maybe you'd like to try VHF/UHF weak-signal operating, but can't put up way to work distant stations on VHF or UHF. If you're chasing awards, you'll pile up the *grid squares* quickly through meteor scatter.

First Contacts

I was all set to go at 0300 UTC. The first sked was easier than I thought. It took about 19 minutes to exchange calls, reports, and final "rogers" with AD4FF. I actually heard him just above the noise on two occasions, indicating some terrestrial propagation was going on as well. Bill is only about 450 miles from me, within range of most well-equipped stations under enhanced conditions.

I calmed my throat with ice-water, turned the antenna toward grid FN30 and relaxed in preparation for the next sked at 0330 UTC with N2OTO. By the start of the next sked, the heat sink on my amplifier had just cooled down. For the next 20 minutes, I called and called, and called, until I was blue in the face. I didn't hear a peep from Rupert. Perhaps east-west wasn't the best path for this shower.

I went back to the calling frequency and tried to pull together another contact. There were a lot of bursts, but most were too short to try a random contact. Just as I was about to pull the plug, I heard Gabor, VE3GBA, and decided to ask about his success. Gabor had already worked two skeds and was on his way to another with KB5IUA in grid EL29. From 0530 to 0600, I listened as Gabor and John exchanged reports. I heard John almost every minute of the sked, sometimes with sustained bursts of up to five seconds. It's surprising how long five seconds sounds during a meteor-scatter contact.

When they finished at 0600, I immediately started calling KB5IUA, hoping he might still be listening. I called for about three minutes before I finally decided to reach him by telephone. I have a directory on my computer that lists stations who are active on VHF and UHF. For each station the directory shows the grid locator, active bands and modes (EME, meteor scatter, etc) and an optional phone number to arrange schedules. John was very accommodating. I thought he wouldn't be too excited to work another station in grid EN82, but he responded, "I love working the rocks! Let's do it!"

We started our sked at 0608 UTC and were done within 10 minutes. On one of the last burns I heard John say, "73, Tom. You got it. That's a good one!" Twelve hundred miles on 2 meters via meteor. Amazing.

During our telephone conversation, John said he would be in grid EL18 in the morning at 1100 UTC. Could we try then? Why not?

With only 3¹/₂ hours of sleep, I was in front of the radio again. The 1100 sked took a little longer, with shorter bursts and weaker signals, but we made it. Instead of getting back in bed, I showered and headed off to work. Felt as if I had a hangover all day, but it was worth it!

a tower. Or, you may live in a valley where hillsides block your signals. The solution? Take to the hills! VHF/UHF antennas are relatively small, and station equipment can be packed up and easily transported. Portable operation, commonly called *hilltopping* or *mountaintopping*, is a favorite activity for many amateurs. During VHF and UHF contests, a station located in a rare grid square is very popular. If you're on a hilltop or mountaintop, you'll have a very competitive signal. See **Table 4-3** for a list of major VHF/UHF contests.

Start by setting up on an easily accessible hill or mountain for an afternoon

How Good Can It Get?

In 1995 hams were wallowing in the trough of a sunspot cycle. The days of almost daily DX on 6 meters seemed long gone, but the veterans knew better than to discount *sporadic E*. Sporadic-E propagation isn't affected by sunspots. Any time, any year, it can appear with surprising results. In the following excerpt from "The World Above 50 MHz" Emil Pocock, W3EP (September 1995 *QST*) chronicles a spectacular 6-meter sporadic-E event that had everyone buzzing for weeks!

Extraordinary Six-Meter Sporadic E!

What a month for sporadic E! Six meters opened to Europe and Africa on 15 days in June, for a total of nearly 60 hours. On two additional dates, North American stations worked the Azores, Madeira, and the Canaries. There have been European openings every June for the past half-dozen years or more, but never have there been so many, or the coverage so widespread. Although the Northeast often had the better of many of these events, stations scattered throughout the eastern half of the country were able to get into Europe.

Many 6-meter operators added new countries in their quest for DXCC—and even the old hands upped their country totals with the appearance of such sought-after stations as EH9IE (Ceuta and Melilla), EH6FB (Balearic Islands), IS0QDV (Sardinia), S59A and S57A (Slovenia), SP6RLA and other Poles, YO2IS and YO7VJ (Romania), and S0RASD (Western Sahara). In all, Americans and Canadians collectively logged more than 30 countries in Europe and North Africa.

As exciting as were the European openings, quite a stir was created from the Caribbean and Central America as well. On at least nine days, propagation favored the south from much of the US. The most sought after stations included TG9AJR (Guatemala), HP2CWB and HP3/KG6UH in Panama, and HR6/W6JKV (Honduras). US operators logged at least 16 different countries from the Caribbean region. Then there were the stations from the north! FP5EK (St Pierre and Miquelon) provided a good deal of excitement both for Americans and Europeans on many days throughout the month. VE8HL (just south of the Arctic circle on Baffin Island) and OX3LX (Greenland) also worked stations throughout the eastern half of the US and Northern Europe.

Table 4-3
Major VHF/UHF Contests

(See QST *magazine for complete details.)*

Contest	Bands	When?
VHF Sweepstakes	50 MHz and up	Varies according to Super Bowl date.
June VHF QSO Party	50 MHz and up	2nd full weekend
CQ Worldwide VHF	50 MHz and up	July
August UHF Contest	222 MHz and up	1st full weekend
September VHF QSO Party	50 MHz and up	2nd full weekend

EME (Moonbounce) with One 4-Element Yagi

By Emil Pocock, W3EP, from "The World Above 50 MHz," August 1995 *QST*

Do you need a huge station to operate Moonbounce? Not necessarily!

Chip Margelli, K7JA/6 (DM03), made a 2-meter EME contact with K5GW on May 5, 1995, at 0108 UTC using a single 4-element Yagi antenna! Chip's 900 W raised his effective radiated power to the same neighborhood as other moderately powered single-Yagi EME stations, but this must be the smallest antenna yet to make an EME contact. This was all the more remarkable because Chip could hear K5GW's signal on the small Yagi and then answered K5GW's CQ—no schedule was involved. Anyone want to try EME with a dipole?

during a contest period. For a first effort, just take along a 2-meter rig. Even if you have an FM-only rig, you can still participate. Use the common simplex frequencies, like 146.52 and 146.55 MHz. If you find that the location "plays," you know where to take your new multimode rig next time!

VHF Contesting

Amateur Radio contests test your ability to work the most stations in different geographical areas on the most bands during the contest period. Contests also give you a chance to evaluate your equipment and antennas, and to compare your results with others. In most VHF/UHF contests, each contact is worth a certain number of points. You multiply your point total by the total number of different grid squares (*multipliers*) to calculate your final score. The only restrictions in these contests are that contacts through repeaters (and satellites) don't count, and the national 2-meter FM calling frequency, 146.52 MHz, is off limits for ARRL contest contacts.

During the first hour or two of a VHF contest, contacts may come fast and furious. At other times, VHF contesting is more like an extended activity hour. VHF contests provide set times during which many other stations are operating. The concentrated activity gives you a chance for many contacts. During a contest, you'll know right away if there's a band opening!

Depending on your location, you

Phil Lee, W6HCC, is an active microwave contester.

may be able to work dozens of different grid squares on several bands, which makes for a high score and lots of fun. If you're interested in awards chasing, you'll also be pleased to know that many hams travel to rare grid squares for contests.

Who Can Enter?

Most VHF/UHF contests are open to any licensed amateur who wants to participate. The ARRL sponsors all the major VHF/UHF contests (see Table 4-3), and specific rules, descriptions of the different categories, as well as entry forms, are available from ARRL Headquarters. You don't have to be an ARRL member to participate in these contests, nor are you required to submit your logs.

VHF/UHF contests feature a variety of categories among which you can choose. For single operators (those operating without assistance), entry classes in the ARRL contests include all-band, single-band, low-power portable, and one for Rovers (those who operate from more than one grid square during the contest). *The ARRL Operating Manual* is a good source of more information on selecting an entry category.

When and Why?

The ARRL VHF contests are held throughout the year, with emphasis on the warmer months to encourage hilltop operation. (Who wants to freeze their toes off contesting from a mountain in subzero cold?) Outside of that, the ARRL VHF/

The SM5FRH 2-meter EME array in Katrineholm, Sweden.

Charles, N2IM, enjoys taking his microwave gear to the nearest hilltops during contests.

Ken, K4DXA, prowls 6 meters, listening for contacts.

UHF contest program is designed to take the best advantage of band openings that usually occur at certain times of the year. For instance, the June VHF QSO Party almost always occurs during periods of excellent sporadic-E propagation, giving you an opportunity to enjoy long-distance contacts on 6 and 2 meters. In fact, the first documented sporadic-E contact on the 222-MHz band was made during a June VHF QSO Party.

As shown in Table 4-3, the major ARRL VHF contests consist of the January Sweepstakes, June and September VHF QSO Parties, August UHF Contest, and the VHF/UHF Spring Sprints. Except for the Sprints, these events encompass many bands each. The January Sweepstakes and June and September QSO Parties are the most popular of them all, and each permit activity on SSB, CW and FM on all amateur frequencies from 50 MHz and up.

The UHF Contest is slightly different from the other contests described so far. The major difference is that only contacts on the 222-MHz and higher bands are allowed.

When to be Where

You'll find lots of random 6 and 2-meter activity during VHF contests. FM is relatively rare on 6 meters in the US, but it's quite common in most areas on 146, 222 and 440 MHz. On SSB, most stations stay near the calling frequencies of 50.125, 50.200, 144.200, 222.100 and 432.100 MHz. On CW, look between 80 and 100 kHz above 50, 144, 222 and 432 MHz. (Six meters offers less CW activity than the other VHF/UHF bands.)

BUILDING YOUR SSB/CW STATION

If you've read this far, I hope you're sold on the idea of trying SSB or CW on

Chinese amateurs have become quite active on 6 meters in recent years. In this photo, Emil Pocock, W3EP (seated, second from right), visited the station of Mars Liu, BG7OH, in Shenzhen and met several of Mars' Amateur Radio friends as well. From left to right, VR2IL; Mars, BG7OH; VR2XMT; W3EP; VR2XRW and VR2PM.

the VHF/UHF bands. Congratulations. You've reached the fun part, when you get to open your checkbook (or grab your credit card) and start shopping for hardware. You have a lot of options available, depending on how much you want to spend.

What's Out There?

Multimode VHF transceivers can be grouped into two classes: home station and mobile/portable. A look at *The ARRL Radio Buyer's Sourcebook* will help you decide what's right for you. It's also a good idea to review recent *QST* Product Reviews when you're selecting equipment.

Many people just getting into VHF settle on multimode, single-band mobile or portable transceivers. These rigs are often less expensive, less complex and more flexible (in terms of power sources and size) than home-station rigs. Some home-station rigs include accessories not usually found in portable and mobile rigs. Serious operators find these accessories, such as preamplifiers, narrow IF filtering and noise blankers helpful when propagation and interference conditions make it hard to hear another station.

Although most VHF multimode transceivers are single-band radios, multi-*band* transceivers have been growing in popularity. Usually aimed at the amateur satellite market, these rigs are also popular among terrestrial operators because of their flexibility. They usually allow you to receive on one band while transmitting on another. These rigs are considerably more expensive than their single-band counterparts, but less expensive than buying separate radios for each band they cover.

Transverters

An alternative to buying one or more VHF transceivers is to buy or build a *transverter* to accompany your HF rig. A transverter takes the RF output from your HF transceiver and uses it to create a signal on a particular VHF or UHF band. The transverter also converts received VHF and UHF signals to HF frequencies. In effect, a transverter turns your HF transceiver into a VHF or UHF transceiver.

Although this equipment requires some effort to interface with an HF rig (except for those made to go with your particular transceiver), the performance and cost savings can be substantial.

And if you don't own an HF transceiver, consider buying a used rig as a "platform" for your transverter. During the solar cycle peak that occurred in the late '80s and early '90s, several manufacturers sold inexpensive 10-meter transceivers. These little rigs were great for working the world with relatively low power (25 W) when the 10-meter band was hot. When the band "cooled" in the mid-'90s, hams started selling these radios at bargain prices. Wise amateurs snapped them up and they soon became the hearts of many VHF and UHF stations!

There are several transverter manufacturers who market to US hams, including . . .

Down East Microwave
954 Rte 519
Frenchtown, NJ 08825
tel 908-996-3584

SSB Electronic
124 Cherrywood Dr
Mountaintop, PA 18707
tel 717-868-5643

Hamtronics
65-Q Moul Rd
Hilton, NY 14468
tel 716-392-9430

VHF/UHF Antennas

We had a pretty robust discussion of VHF and UHF antennas in Chapter 2. The same principles apply with one exception: antennas for SSB and CW operating must be *horizontally polarized*. That is, the parts of the antennas that radiate and receive energy must be *parallel* to the ground.

Why?

Because that's the way everyone else does it! Seriously, it's more than a matter of group consensus. If your antenna is vertically polarized and the other fellow is using a horizontally polarized antenna, his signal will be substantially weaker than it should be on your end, and vice versa.

The loss caused by mismatched polarization is most noticeable during line-

If the microwave bug really bites, you may eventually end up with an antenna farm like the one owned by Tom, WA8WZG.

of-sight contacts. If your signal is bouncing off a meteor trail or a layer of the atmosphere, polarization doesn't mean much. What goes up vertical, for example, comes down every which way!

Because we're dealing with "weaker" signals, directional antennas are best. You want to focus your signal as much as possible. There is no law that says you can't use an omnidirectional antenna on SSB or CW, but your effective range will be very limited.

Directional antennas include the venerable Yagi and the quad. The Yagi is the most common directive antenna. Yagi antennas are commercially available with three to at least 33 elements. The quad uses loop elements instead of wire or rod elements.

If you want multiband performance in a single directional antenna, consider the *log-periodic dipole array* (LPDA), usually referred to simply as a *log periodic*. A log periodic beam antenna covers several bands. The penalty for this frequency coverage is significantly lower gain than can be achieved with a single-band Yagi or quad. Log periodics are also more mechanically complex than Yagis and quads. On the other hand, the convenience of having coverage of so many bands with only one antenna and feed line is very attractive, especially for portable operation.

How Do I Choose?

There's even more variety in antennas than in equipment for the VHF bands. Unless your local dealer carries all the major brands, you'll do most of your shopping without actually seeing all available antennas up close. That's why it's a good idea to get catalogs from major retailers who advertise in *QST*, which show at least the major specifications for the antennas they carry. Also, read *QST* Product Reviews, and ask your friends what they're using.

Feed Lines: The Weak Link

When you install your antennas, you'll need to connect them to your radios via feed lines. No surprise so far, right? What makes this subject worth discussing here is *loss*.

If you missed the discussion of feed line loss in Chapter 2, go back and read it. Choosing the right coaxial cable is important for FM operating . . . and it's downright *critical* for SSB and CW! Always choose the lowest-loss cable you can afford.

Don't blow your savings on a nice transceiver and a whiz-bang antenna only to scrimp on the coax. You'll be very sorry. Many hams have tried their hands at "weak signal" operating, but they were handicapped from the start because they bought coax that was way too lossy for the installation. Not only did they lose precious transmit energy, their received signals were watered down as well.

THE CHALLENGE AND THE REWARD

There's no question that it's easier to get on the air with FM than SSB or CW. With FM, it may be a matter of simply buying a hand-held transceiver and talking through your local repeater. SSB and CW take a little more effort, but the reward is considerable!

As a "weak-signal" operator, you'll enjoy contacts over distances that FM enthusiasts can only achieve through complex linked-repeater systems. Best of all, you'll experience the true magic of VHF operating. As you sharpen your skills, you'll be able to predict when band openings are about to take place. By listening to the distant signals, you'll know which propagation mode is active and how to use it to your advantage.

VHF operating will challenge you every day. DX stations sometimes appear when you least expect them—and disappear just as suddenly. Wait until the day when you turn on your equipment and hear a flood of distant CW and SSB signals. The excitement will be electrifying and you'll know in that moment what you've guessed all along: there is much more to VHF than FM!

5 | Where No Ham Has Gone Before

It's a long way from Starfleet, but there is an armada of Amateur Radio spacecraft in orbit above our planet at this very moment. (Quick, ma! Grab the telescope!) When this book went to press, "Hamfleet" was comprised of more than 15 satellites.

Believe it or not, amateur satellites have been in orbit since the early '60s. Even before astronaut John Glenn made his historic flight, OSCAR 1 (*O*rbiting *S*atellite *C*arrying *A*mateur *R*adio) was circling the Earth, transmitting "HI" in CW.

Today you can choose from a variety of extremely sophisticated amateur satellites. You can even communicate with the crew aboard the International Space Station. What may surprise you more than anything else, however, is the ease with which you can access most of these satellites. It seems to be one of the best-kept secrets in Amateur Radio!

FINDING THE SATELLITES

Before you can use any amateur satellite, you need to find it—or at least be able to predict where it will be at a given time. This is a little more complicated than it sounds. Now don't get intimidated! "Complicated" doesn't mean "difficult." It just takes a few extra steps. Allow me to explain . . .

As of this date, all Amateur Radio satellites are in nongeostationary orbits. This simply means that the satellites are not in fixed positions in the sky from our perspective here on Earth. They are like tiny moons, rising and setting over your local horizon.

If a low-altitude satellite passes over your house at 9 AM, it will make a complete orbit and return to the same spot about 100 minutes later . . . or will it? It might, *if the Earth remained perfectly still*. But the Earth is turning at a pretty good clip and it takes you, your house, your car and everything else in your neighborhood right along with it. By the time the satellite completes its orbit, you'll be several hundred miles east of where you were before. So, the satellite won't appear over your house again; it will be screaming over someone else's home.

The problem gets stickier when you talk about satellites that have unusual orbits. For example, some satellites zoom in very close to the Earth, then go flying far out into space. (Imagine a kind of cosmic yo-yo.)

Orbital Elements

How can you know when a satellite is about to make an appearance in your neighborhood? To answer that question you need to know the satellite's *orbital elements* (also called Keplerian elements).

An orbital element set is merely a collection of numbers that describes the movement of an object in space. By feeding the numbers to a computer program, you can determine exactly where a satellite is (or will be) at any time. Before the heyday of PCs, hams did their orbital calculations manually. You can learn a lot about orbital mechanics from the manual method, but few prefer to do it that way these days. Grabbing a set of orbital elements is as easy as jumping on the Web and going to the AMSAT-NA site at **www.amsat.org**. If all else fails, there is probably someone in your area who has access to the elements. Ask around at your next club meeting. With a couple of exceptions, you only need to update your elements every few months.

Figure 5-1—*Nova* **satellite-tracking software displays the footprint of AMSAT-OSCAR 40.**

Feeding the Software

There are many *satellite-tracking* programs on the market that will take your orbital elements and magically produce satellite schedules. Some will read the elements indirectly; others require you to type them in manually.

Among other things, these programs tell you when satellites will appear above your local horizon and how high they will rise in the sky (their elevation). When working satellites, the higher the elevation the better. Higher elevation means less distance between you and the satellite with less signal loss from atmospheric absorption.

Some programs also display detailed maps (**Figure 5-1**) showing the satellite's *footprint* over the ground. The footprint is the area on the Earth that is within direct line-of-sight range of the satellite. You'll find a wealth of satellite tracking software at the AMSAT Web site. You may also find less sophisticated tracking software that is either freeware or shareware on the Web. Try a "search engine" such as Google (**www.google.com**) and I bet you'll find quite a bit of software available for downloading.

OSCARS 14 AND 27: FM REPEATERS IN ORBIT

OSCARs 14 and 27 are basically FM repeaters in outer space. You may also see them referred to by their abbreviated names: UO-14 and AO-27 respectively. These satellites only transmit at about 1 to 4 W output, but that's plenty when you consider the height of their antennas!

The FM birds are popular among hams who own dual-band FM transceivers— especially if the 440-MHz section offers receive and transmit coverage down to 436 MHz. (Not all dual-banders have this feature. Check your transceiver manual.) To make contacts through the FM repeater satellites you transmit on one band and receive on another (see the frequencies in **Table 5-1**). You don't need beam antennas to work these satellites. Some hams have even managed to make contacts using hand-held rigs and rubber-duck antennas! Of course, the better your antenna, the better your odds of success—especially when the satellites are crowded.

Because only one station at a time can talk through the satellites, contacts tend to be *very* short! These low-earth-orbiting (LEO) birds are only in view for about 20 minutes at best, so lots of hams jump on the air and try to make contacts.

The **AMRAD-OSCAR 27 satellite about a month prior to launch.**

If everyone is polite and patient, you'll be able to make a quick contact, tell the other station where you are, and then say "good-bye." Human nature being what it is, what you'll probably hear instead is a distorted squeal from the satellites as too many stations transmit at once. In this situation "survival of the fittest" rules the radio jungle. He who has the most power and the biggest antenna will have the clearest signal. If you're lucky, the big-signal station will appoint him or herself as a kind of "net control" and direct the flow of contacts during the pass. If not, it's a radio slugfest!

Table 5-1

Amateur Satellites: Frequencies and Modes

Note: OSCAR 40 frequencies are shown in Table 5-2.

Satellite	Uplink (MHz)	Downlink (MHz)
SSB/CW		
AMSAT-OSCAR 10	435.027 - 435.179	145.825 - 145.977
Fuji-OSCARs 20 and 29	145.900 - 146.000	435.800 - 435.900
RS-13 (Mode K)	21.260 - 21.300	29.460 - 29.500
RS-13 (Mode A)	145.960 - 146.000	29.460 - 29.500
Digital—1200 bit/s		
(FM FSK uplink, PSK downlink)		
AMSAT-OSCAR 16	145.90, .92, .94, .96	437.05/437.026
LUSAT-OSCAR 19	145.84, .86, .88, .90	437.126/437.15
ITAMSAT-OSCAR 26	145.875, .900, .925, .950	435.870
Packet—9600 bit/s		
(FM FSK uplink and downlink.)		
UoSAT-OSCAR 22	145.900, .975	435.120
KITSAT-OSCAR 23	145.85, .90	435.175
KITSAT-OSCAR 25	145.87, .98	436.50
Thai-OSCAR 31	145.925	436.925
UoSAT-OSCAR 36	145.960	437.025, 437.400
FM Voice		
AMRAD-OSCAR 27	145.850	436.795
UoSAT-OSCAR 14	145.975	435.070
International Space Station	145.800 (simplex)	

THE INTERNATIONAL SPACE STATION (ISS)

An Amateur Radio station was activated aboard the International Space Station by its first crew in late 2000. The ISS station sports a powerful signal that you'll usually hear on 2-meter FM. At the time this book was written, Amateur Radio contacts with the ISS crew were somewhat sporadic because they still were intensely involved in setting up the new facility. Even so, some hams have been lucky enough to snag voice contacts with the crew on 145.800 MHz and there have been several scheduled contacts with schools as well. Ham activity from the International Space Station is certain to increase over the coming years.

Working the ISS on voice is very similar to working a DX pileup. You sit with microphone in hand and wait until you hear the crewmember complete an exchange. At that moment you key the mike and say your call sign. Now listen. No response? Call again quickly! Keep trying until you hear him calling you or someone else.

The first expedition crew sent to the International Space Station when it was ready for occupation in 2000 included two Amateur Radio operators. Left to right: Sergei Krikalev, U5MIR; William Shepherd, KD5GSL and Yuri Gidzenko.

I've heard of hams working the ISS while mobile and some claim to have worked the ISS with hand-helds. As you might imagine, ISS QSL cards are highly prized!

The major problem with working the ISS is its erratic schedule. The crew has many daily assignments and is not always able to find the time to operate their amateur station. They are sometimes forced to turn off their equipment altogether to avoid interference to other systems during critical tests.

Another problem concerns the International Space Station's orbit. The space station travels at a relatively low altitude, so it's always subject to a significant amount of atmospheric drag. If it didn't occasionally "boost" to a higher orbit, the station would reenter the atmosphere and be destroyed. Every time the ISS fires its rocket engines to adjust its orbit, a revised set of orbital elements must be distributed. If you want to try your luck with the ISS, plan to update your elements for the space station as often as possible.

RS (RADIO SPUTNIK) 13

The RS-13 satellite is completely different from the FM repeater satellites or the International Space Station. RS-13 is basically orbiting SSB/CW *repeater* riding piggyback on a larger navigational satellite. What makes RS-13 even more unusual is that it is equipped with a device known as a *linear transponder*.

Linear Transponders

Earthbound repeaters listen on one frequency and repeat what they hear on another. So do the FM repeater satellites. But imagine what would happen if your local repeater could retransmit everything it heard on an entire group of frequencies—not just one conversation, but *several at once*? This is exactly the function of a linear transponder.

When this book was written, RS-13 was listening to a portion of the 2-meter band and retransmitting on the 10-meter band (see **Figure 5-2**). This type of operation is known as *Mode A*. On occasion the ground controllers switch RS-13 to listening on 15 meters and retransmitting on 10 meters. This type of operation is known as *Mode K*. The range between the highest and lowest uplink (or downlink) frequencies is known as the transponder's *passband*. See the RS uplink and downlink passbands in Table 5-1.

Not only do linear transponders repeat everything they hear on their uplink passbands, they do so very faithfully. CW is retransmitted as CW; SSB as SSB. FM voice transmissions are strongly discouraged since their broad signals

Figure 5-2—A linear transponder acts much like a repeater, except that it relays an entire group of signals at once, not just one signal at a time. This is an illustration of the linear transponder aboard RS-13. It listens on the 2-meter band and repeats everything it hears on a portion of 10 meters.

RS-13 Worked All States

By R. A. Peschka, K7QXG

There was a time when I thought satellite operations were reserved for the technically elite of our Amateur Radio society. That misguided perception prevailed until April 1993 when I overheard an interesting conversation between Roger, N4ZC, and a station in Puerto Rico. N4ZC was encouraging the KP4 to try a contact on the next pass of the RS-13 satellite using mode K (mode K means you transmit on 21 MHz and listen on 29 MHz). Their conversation intrigued me so much I decided to give it a try myself. A short time later I had made contact with N4ZC via the RS-13 satellite and my space odyssey began.

Sharpening My Skills for WAS

N4ZC proved to be a great help. He sent me a list of the times when I could expect the "bird" to pass within my range. It wasn't long before it became clear that the use of a good computer program would go a long way toward making these operations even more fun and rewarding. I joined AMSAT; acquired a good program for my PC; and, now charged with excitement, was active on the satellite. After a month or so, it became apparent that the challenge of chasing my Worked All States (WAS) award on RS-13 would be an excellent way to tune up my operating skills, which had become quite rusty.

At first I evaluated each projected pass of the satellite. I designed a precise strategy; but I threw all my good intentions to the wind when the thrill of the hunt took over. I'm sure I missed many golden opportunities to contact some of those hard-to-reach stations.

I began making progress when I took my methods seriously. After every pass I evaluated the results to see what worked—and what didn't. Part of the trick was becoming intimately familiar with every little knob and dial on my transceiver. I began to spend far more time listening than transmitting, and learned how to compensate for noise, Doppler shift, and all the other nuances encountered when working the "bird." I honed my skills with practice and the states fell one by one.

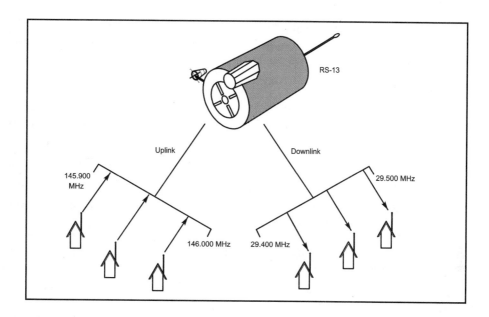

145.900 MHz

Uplink

146.000 MHz

Downlink

29.500 MHz

29.400 MHz

RS-13

Down to the Last States

After nearly a year, I needed only five states to reach my goal: South Dakota, Wyoming, Montana, Delaware and Maine. A good friend and fellow satellite hunter suggested Clarence, N7RPC in Wyoming. I fired off a letter to Clarence, seeking a schedule on RS-13. The letter resulted in a telephone call, a bit of preparation and a QSO attempt. After some momentary confusion, followed by another quick telephone call, we made contact. Meanwhile, by pure luck, I managed to find WØIT in South Dakota.

Clarence, in turn, suggested Ken, WG7G, in Montana as a possibility to move up one more notch. A letter was sent to Ken, but before his reply arrived he found me on the satellite! With Montana in the bag I had a total of 48 states! Only Delaware and Maine remained elusive.

As I chased the states I renewed operating skills long forgotten, not the least of which was patience and courtesy. I once again became familiar with the nuances of having pens run out of ink at the precisely wrong moment; of having two too many thumbs, and of drinking too much coffee before the "window" opened to the usual flurry of activity.

The search for Delaware and Maine continued. Finally, I heard Gene, NY3C, in Delaware, but the satellite sank below the horizon before I could make contact. It was letter writing time again. Gene really had the true ham spirit. He called me by telephone and we arranged a list of possible schedules. We made contact on the very first try. That left only Maine.

I had sent letters to a couple of stations in Maine, but never received a response. NY3C stepped into the breach again and spread the word of my quest on various nets. Two days later, on my favorite CW frequency, I was called by W1OO in Maine!

About RS-13

This amateur satellite is actually part of a large Russian COSMOS navigational satellite launched in 1988. It was assembled at the Tsiolkovskiy Museum for the History of Cosmonautics in Kaluga (about 180 km southwest of Moscow).

Although it is capable of operating in a number of modes, RS-13 only functions in mode K at this time. The frequencies are shown in Table 5-1.

occupy an enormous chunk of the downlink passband. Not only would this limit the number of stations that could use the satellite, it would place a severe drain on transponder power.

Working RS-13

For RS-13 all you need is an HF transceiver or receiver capable of listening around 29.5 MHz. and a 2-meter multimode transceiver (a radio that can transmit in CW or SSB).

If you find that RS-13 is operating in Mode K, there is one "legal" catch involved. Take another look at the RS-13 15-meter uplink passband in Table 5-1. If you have a chart of US amateur HF frequencies, you'll quickly discover that you must have an General license to transmit within that passband. On the other hand, Technician hams should know that it is perfectly legal for RS-13 to relay their signals in Mode A to the upper portion of the 10-meter band—even though they do not have privileges there. After all, they are not the ones transmitting on 10 meters; RS-13 is!

Antennas

Elaborate antennas are definitely not required to work RS-13. A wire dipole is fine for receiving the 10-meter downlink signal. By the same token, a basic 2-meter antenna such as a ground plane or J-pole is fine for an uplink to RS-13. In terms of power, 20 to 30 W on the 2-meter uplink seems to work well.

With the wide separation between uplink and downlink frequencies, you can work the RS-13 satellite in *full duplex* if you have a separate 10-meter receiver. When operating full duplex, you can hear your own signal on the satellite down-link *while you're transmitting*! When you're listening to your own signal through the satellite, you'll notice something odd right away—your signal keeps drifting downward in frequency. That's why you need to tweak the VFO on your uplink transmitter to keep your signal on frequency. All these acrobatics are caused by a pesky problem known as *Doppler shift*.

The Mystery of the Shifting Signal

Doppler shift is caused by the difference in relative motion between you and another object—a satellite in this case. As the satellite moves toward you, the signal frequencies in the downlink passband gradually *decrease*. It's the same effect you hear when a locomotive approaches a crossing. The audio frequency of the horn decreases as the train moves toward you, then away. The higher the frequency, the more pronounced the effect. Doppler, for example, is much more noticeable on 435-MHz signals than on 29-MHz signals.

Dealing with Doppler is not that difficult once you get used to it. The standard rule of thumb is to change the frequency of your *uplink* to compensate, while leaving your downlink frequency the same. As you listen, you hear your signal shifting downward. Don't retune your receiver! Instead, tweak your *transmitter*

VFO as much as it takes to keep your voice sounding normal. We'll discuss this technique in more detail later.

RS-13 Operating Techniques

Since the satellites are available for a relatively short time, contacts tend to be rather brief. CW operators congregate in the lower half of the transponder passband while SSB operators occupy the upper half.

If you hear someone calling CQ on SSB, note the downlink frequency and quickly tune your transmitter accordingly. As you answer the call, adjust your transmitter until your voice is clear and audible on the downlink. (*If* you're operating full duplex.) You can even do this while he is still calling CQ. I've heard some SSB operators adjusting their uplink frequency and saying, "Test, test, test..." By using this method they're assured of being on-frequency and ready to respond when the other station stops calling.

Answering a CW call is just as easy. As soon as you copy the call sign, tune your transmitter to the proper frequency and start sending a series of dits. Listen on the downlink and adjust your transmitter until you hear the tone of your CW signal roughly matching the tone of the station sending CQ.

A Japanese rocket carries Fuji-OSCAR 20 to orbit.

FUJI-OSCARS 20 AND 29

Like RS-13, OSCARs 20 and 29 feature linear transponders. They repeat all conversations taking place within their uplink passbands. Most stations working through OSCARs 20 and 29 use SSB, although some CW signals are heard as well. With their relatively high orbits, OSCARs 20 and 29 offer excellent coverage. For example, hams on the East Coast of the United States can easily work Western Europe when OSCARs 20 or 29 are above the mid-Atlantic.

You'll need a 2-meter multimode transceiver to send signals to OSCARs 20 and 29 and a 440-MHz multimode transceiver (or simply a 440-MHz SSB receiver) to hear the downlink (see **Figure 5-3**). This is a substantial investment for most amateurs. In addition, beam antennas are best for these birds, along with an azimuth/elevation antenna rotator to provide horizon-to-horizon tracking. (More money!)

As a result, you don't hear many signals on OSCARs 20 and 29. Even so, investing in an OSCARs 20 and 29 station will pay off down the road when it comes time to work more sophisticated birds.

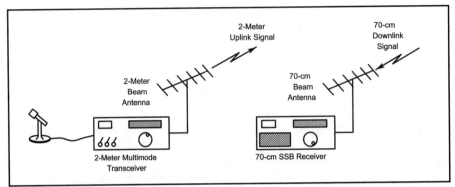

Figure 5-3—A station equipped for OSCARs 20 and 29 requires a 2-meter multimode transceiver, a 435-MHz SSB receiver, and beam antennas for the uplink and downlink signals.

THE DIGITAL SATELLITES

If you enjoy digital hamming, you'll love the digital satellites. Several satellites comprise the digital amateur fleet: AMSAT-OSCAR 16, Lusat-OSCAR 19, UoSat-OSCAR 22, KITSAT-OSCAR 23, KITSAT-OSCAR 25, Thai-OSCAR 31, ITAMSAT-OSCAR 26 and UoSat-OSCAR 36.

Most digital satellites work like temporary mailboxes in space. You upload a message or some other file to a digital satellite and it is stored for a time (days or weeks) until someone else—possibly on the other side of the world—downloads it. Neat, isn't it?

Which Digital Satellite is Best?

You can divide the digital satellites into two types: The 1200 and 9600-baud satellites. OSCARs 16, 19 and 26 are 1200-baud digital satellites. You transmit data to them on 2-meter FM and receive on 437-MHz SSB. All the rest are 9600-baud digital satellites. You send data to them on 2-meter FM and receive on 435-MHz FM.

So which digital satellites are best for beginners? There's no easy answer to that question. You can use any 2-meter FM transceiver to send data to a 1200-baud digital satellite, but getting your hands on a 435-MHz SSB receiver (or transceiver) could put a substantial dent in your bank account (see **Figure 5-4**). In addition, you need a special PSK (*phase-shift keying*) terminal node controller (TNC). These little boxes are not common and could set you back about $250.

So the 9600-baud digital satellites are best for the newbie, right? Not so fast. It's true that you don't need a special packet TNC. Any of the affordable 9600-baud TNCs will do the job (see **Figure 5-5**). The catch is that not any FM transceiver is usable for 9600-baud packet. Read the discussion of 9600-baud packet in Chapter 2 and you'll see what I mean. *Both* the 2-meter *and* 440-MHz FM radios must be capable of handling 9600-baud signals. And not all 440-MHz FM rigs can receive down to 435 MHz.

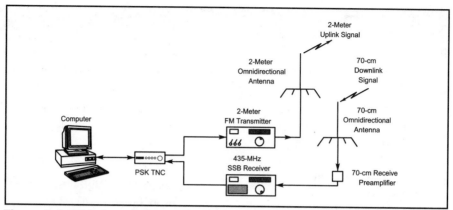

Figure 5-4—This is a diagram of a typical 1200-baud digital satellite station. Notice the special PSK TNC and the 435-MHz SSB receiver. For improved performance, substitute beam antennas.

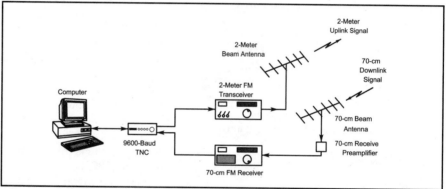

Figure 5-5—A 9600-baud digital satellite station requires less specialized equipment, but the radios must be capable of handling 9600-baud data signals.

Broadcasting Data

Despite the huge amounts of data that can be captured during a pass, there is considerable competition among ground stations about exactly *which* data the satellite should receive or send! There are typically two or three dozen stations within a satellite's roving footprint, all making their various requests. If you think this sounds like a recipe for chaos, you're right.

The digital satellites produce order out of anarchy by creating two *queues* (waiting lines)—one for uploading and another for downloading. The upload queue can accommodate two stations and the download queue can take as many as 20. Once the satellite admits a ground station into the queue for downloading, the

Budgeting Your Satellite Station

The following price estimates are based on *new* equipment. You can expect to save 50% or more if you explore the used-equipment market—and I encourage you to do so.

Note that many of the same components can be used for more than one satellite. The list also does not include other items such as coaxial cable.

The International Space Station
- 2-meter FM transceiver ($300)
- Omnidirectional antenna ($30)

RS-13:
- HF SSB/CW transceiver ($600)
- 2-meter transceiver ($600)
- Wire antennas ($50)

OSCARs 14 and 27
- Dual-band FM transceiver (must be able to receive down to 436 MHz) ($400)
- Dual-band omnidirectional antenna ($100)

OSCARs 20 and 29
- 2-meter multimode transceiver ($600)
- 440-MHz multimode receiver or transceiver ($500)
- 440-MHz receive preamplifier—*installed at the antenna* ($150)
- Beam antennas for 2-meters and 440 MHz ($200)
- Azimuth/elevation antenna rotator ($500)

1200-baud Digital Satellites
- PSK satellite TNC ($300)

- 2-meter FM transceiver ($500)
- 440-MHz multimode receiver or transceiver ($500)
- Dual-band omnidirectional antenna ($100)

9600-baud Digital Satellites
- 9600-baud TNC ($200)
- 2-meter FM transceiver with 9600-baud capability ($500)
- 440-MHz FM transceiver with 9600-baud capability ($500)
- Dual-band omnidirectional antenna ($100)

OSCARs 10 or 40
- 2-meter/440-MHz multimode transceiver ($1300)
- 100-W 440-MHz power amplifier and power supply ($500)
- 2-meter receive preamplifier—*installed at the antenna* ($150)
- Beam antennas for 2 meters and 440 MHz ($300)
- Azimuth/elevation antenna rotator ($500)

---

station moves forward in the line until it reaches the front, whereupon the satellite services the request for several seconds.

For example, let's say that OSCAR 16 just accepted me, WB8IMY, into the download queue. I want to grab a particular file from the bird, but I have to wait my turn. OSCAR 16 lets me know where I stand by sending an "announcement" that I see on my monitor. It might look like this:

WB8ISZ AA3YL KD3GLS WB8IMY

WB8ISZ is at the head of the line. The satellite will send him a chunk of data, then move him to the rear.

AA3YL KD3GLS WB8IMY WB8ISZ

Now there are only two stations ahead of me. When I reach the beginning of the line, I'll get my share of "attention" from the satellite.

Unless the file you want is small, you won't get it all in one shot. If the satellite disappears over the horizon before you receive the complete file, there's no need to worry. Your digital satellite software "remembers" which parts of the file you still need from

WISP software in action as it communicates with the KITSAT-OSCAR 23 packet satellite. In the lower left portion of the screen you can see the list of stations waiting in the queue.

the bird. When it appears again, your software can request that these "holes" be filled.

And while all of this is going on, *you're receiving data that other stations have requested!* That's right. Not only do you get the file you wanted, you also receive a large portion of the data that other hams have requested. You may receive a number of messages and files without transmitting a single watt of RF. All you have to do is listen. That's why they call it "broadcast" protocol.

Station Software

You must run specialized software on your station PC if you're going to enjoy any success with the digital satellites. If you're running Microsoft *Windows* on your PC, you'll want to use *WISP*. This software package is available from AMSAT at the address shown in the Info Guide section of this book.

OSCAR 40 AND OSCAR 10—THE DX SATELLITES

Throughout this chapter we've been talking about satellites that travel in low-Earth orbits. The advantages of these satellites are obvious: you can access them with relatively low power and very meager antennas. On the other hand, they are available only for brief periods of time and the opportunities for DX (long-range communication) are almost nonexistent.

Fortunately, there are two satellites that travel in high, elliptical orbits and they're both DX powerhouses: AMSAT-OSCAR 40 and AMSAT-OSCAR 10.

As you'll see in **Figure 5-6**, both satellites have orbits that act like slingshots, shooting them out to altitudes of more than 30,000 km. At the high points of their orbits, they seem to be nearly motionless from our perspective here on Earth. While a certain amount of antenna aiming is required, very little additional movement is necessary once the antennas are in their proper positions. From their high vantage points, OSCARs 10 and 40 "see" a great deal of the Earth. This opens a window to DX contacts on a regular basis!

The downside to having such a high-altitude orbit is that more transmitted power is needed to access the satellites and a weaker signal is received. You'll need high-gain, directional antennas to operate OSCARs 10 or 40. The more gain you have at the antenna, the less power will be required at the transmitter. You'll need to be able to rotate the antennas vertically *and* horizontally (elevation and azimuth).

Mode UV (70-cm uplink/2-meter downlink) is the most popular mode on OSCAR 10. (See the frequencies listed in Table 5-1.) A 2-meter SSB/CW receiver is required for the downlink and a similar 70-cm transmitter is necessary for the uplink. Considering the weak signals, a 2-meter mast-mounted preamplifier is also a worthy addition to your station.

Like the RS satellites, OSCARs 10 and 40 employ linear transponders to "repeat" many signals at once. SSB and CW are the modes of choice, although conversations tend to be longer and more relaxed. With these satellites you don't have to worry too much about losing the signal in the middle of your conversation!

OSCAR 10 is in a stable orbit, but it has had a rough life. High levels of radiation have hammered the poor satellite and they've taken their toll. It's out of control in the sense that it doesn't respond to orders from the command stations. (They gave up long ago.) It simply cruises through space, repeating whatever signals it happens to pick up. On some days, OSCAR 10 is terrific. It will sound every bit as good as OSCAR 40. On other days, you can barely hear it. Eventually

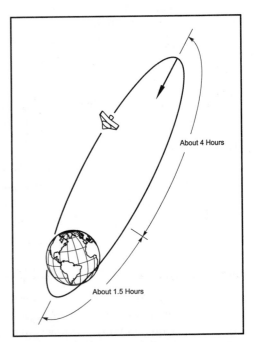

Figure 5-6—OSCARs 10 and 40 travel in high, elliptical orbits that shoot the spacecraft far out into space.

Table 5-2

Transponder Frequency Band Plan for AMSAT-OSCAR 40

Uplink Frequencies

Band	Digital	Analog Passband
15 m	none	21.210 - 21.250 MHz
12 m	none	24.920 - 24.960 MHz
2 m	145.800 - 145.840 MHz	145.840 - 145.990 MHz
70 cm	435.300 - 435.550 MHz	435.550 - 435.800 MHz
23 cm(1)	1269.000 - 1269.250 MHz	1269.250 - 1269.500 MHz
23 cm(2)	1268.075 - 1268.325 MHz	1268.325 - 1268.575 MHz
13 cm(1)	2400.100 - 2400.350 MHz	2400.350 - 2400.600 MHz
13 cm(2)	2446.200 - 2446.450 MHz	2446.450 - 2446.700 MHz
6 cm	5668.300 - 5668.550 MHz	5668.550 - 5668.800 MHz

Downlink Frequencies

Band	Digital	Analog Passband
2 m	145.955 - 145.990 MHz	145.805 - 145.955 MHz
70 cm	435.900 - 436.200 MHz	435.475 - 435.725 MHz
13 cm(1)	2400.650 - 2400.950 MHz	2400.225 - 2400.475 MHz
13 cm(2)	2401.650 - 2401.950 MHz	2401.225 - 2401.475 MHz
3 cm	10451.450 - 10451.750 MHz	10451.025 - 10451.275 MHz
1.5 cm	24048.450 - 24048.750 MHz	24048.025 - 24048.275 MHz

Telemetry Beacons

Band	General Beacon (GB)	Middle Beacon (MB)	Engineering Beacon (EB)
2 m	none	145.880 MHz	none
70 cm	435.450 MHz	435.600 MHz	435.850 MHz
13 cm(1)	2400.200 MHz	2400.350 MHz	2400.600 MHz
13 cm(2)	2401.200 MHz	2401.350 MHz	2401.600 MHz
3 cm	10451.000 MHz	10451.150 MHz	10451.400 MHz
1.5 cm	24048.000 MHz	24048.150 MHz	24048.400 MHz

the bird will go silent for good, but no one knows when that might happen.

The Flagship—OSCAR 40

[Editor's note: As this is written, OSCAR 40's capabilities have not been fully determined.]

OSCAR 40 is, without a doubt, the largest, most expensive Amateur Radio satellite ever created. It's an international effort with components supplied by amateur satellite organizations and individuals throughout the world. Those sections that resemble wings are actually solar panels. They unfold to provide power to the satellite after it reaches orbit.

OSCAR 40 channels this power to run an incredible array of transmitters and receivers on frequencies from 21 MHz to 24 GHz. The RF output of its 2-meter transmitter alone is about 200 W. OSCAR 40's 2-meter antenna offers an effective

Satellite Antennas

As with any Amateur Radio station, your antennas determine how well you'll communicate. Connecting an expensive radio to a lousy antenna is a sure-fire formula for frustration!

Directional (beam) antennas such as Yagis or quads are best for any satellite station. They concentrate your transmitted and received energy, allowing you to more easily bridge that multihundred or multi*thousand* mile gap between you and the satellite.

The main problem with beam antennas, however, is cost and space. Beam antennas are not cheap, and you'll need expensive azimuth/elevation rotators to use them properly. Az/el rotators spin the antennas side to side as well as up and down. A new unit typically costs $500. Beam antennas and rotators also require space on your roof or wherever. Not every ham has the room to install such a system.

So, some satellite-active hams must compromise and use omnidirectional antennas.

Classic cross-polarized Yagi antennas like these are the ultimate performers at any satellite station. As you can see from this photograph, they don't necessarily need to be installed on rooftops, either! The larger beam in the foreground is used for the 2-meter downlink for OSCAR 40. The small beam in the background is the 70-cm uplink antenna.

The *eggbeater* is an omnidirectional antenna that creates a circular radiation pattern overhead. This is ideal for satellite applications.

A close-up view of an azimuth/elevation antenna rotator. This device will move your antennas from side to side (azimuth) as well as up and down (elevation). The small boxes immediately below the rotator are receive preamplifiers.

These antennas are more affordable, fit into relatively tight spaces and do not require rotators. On the other hand, they have no signal-concentrating ability.

The Importance of Coax

Regardless of which antenna system you select, be sure to connect them to your radios with the best coaxial cable you can afford. Don't cut corners here! As a rule of thumb, you can use Belden 9913 coax (or equivalent) for all your satellite antennas. You may also find *LNR* coax that has similar low-loss characteristics. Either 9913 or one of the LNR varieties are good choices for the satellite stations we've discussed in this chapter.

Polarization?

In Chapter 2 I said that antennas for FM communication must be *vertically* polarized. In Chapter 3 I stated that antennas for SSB or CW on the VHF/UHF bands must be *horizontally* polarized. What type of polarization is best for satellite work?

For most satellites, antenna polarization is not quite so critical. If you're using an omnidirectional antenna, vertical polarization is fine. You may encounter a few omni antennas that offer *circular* polarization (such as the so-called "eggbeater" designs). These are even better for satellites, but they cost substantially more.

If you browse through the ham-equipment catalogs, you'll find *cross-polarized* beam antennas for satellite stations. These are basically Yagi antennas with both horizontal *and* vertical elements. A remote-control relay allows you to select one set of elements or the other.

Do you need cross-polarized beams? For the high-altitude birds such as OSCARs 40 and 10, a cross-polarized beam can give you a decided signal advantage. For the other satellites, their value is questionable. A set of vertically polarized beams often works just as well.

radiated power (ERP) of 180 W. The superior 2-meter antennas aboard OSCAR 40 are capable of yielding an ERP of up to *2500 W!*

What does this mean to you? It means that you don't need the large multielement beam antennas you're accustomed to seeing on most OSCAR 10 stations. Depending on how sensitive your receiver is, you may not even need a mast-mounted receive preamp.

OSCAR 40 really shines on the bands above 1 GHz. Let's face it, the future of amateur satellite communication is microwave. On the microwave bands you can get superb gain from tiny antennas. Noise levels are extremely low, too. OSCAR 40's microwave ability make it possible to install your satellite station in a cramped attic with room to spare—or on the balcony of your apartment building. Your neighbors won't even know your station exists.

But microwave equipment is difficult to use or build, right? Wrong. The days of microwave "plumbing" projects are behind us. Now we have *transverters* that will take 2 or 10-meter signals and turn them into 1.2, 2.4, 10 or 24-GHz satellite uplinks. You can buy these transverters or build them. (Yes, they're available as kits.)

AMSAT-OSCAR 40 is the largest, most complex Amateur Radio satellite ever launched.

The 2400-144RX receive controller from Down East Microwave will convert the 2.4 GHz signals from OSCAR 40 to 2 meters (144 MHz).

While SSB and CW are the primary modes of communication through OSCAR 40 (sorry, the power budget won't permit the use of FM), the satellite also functions as a relay for digital signals. Packeteers can take advantage of OSCAR 40's *RUDAK* system to transfer data over vast distances. The nominal data rate will be 9600 bit/s, but RUDAK is designed to operate at much higher rates. Some see the day when OSCAR 40 may be the linchpin in a network of packet satellite gateways that will communicate at 56 kbit/s or beyond.

6 | The Camera Never Lies

"Hams . . . and the people who share their secret desires! It's all coming up today on the AA6XYZ show!"

(Switch to wide-angle pan. Show audience of neighborhood kids, dog, goldfish and petulant spouse.)

(Dissolve to call sign logo.)

(Fade to black.)

Okay, not all hams make "productions" out of their amateur television transmissions, but it's tempting, isn't it? After all, this is *your* television station. You call the shots.

I bet the idea of operating a television station from your home probably makes you want to hide your checkbook and credit cards in a safe place. That's understandable. Amateur television (ATV) must surely be the most expensive aspect of our multifaceted hobby, right?

Uh . . . no.

Depending on how fancy you want to get, you can put your own ATV station on the air for hundreds of dollars *less* than the cost of a new transceiver.

FAST SCAN TV

Fast scan TV (or FSTV) is the same type of television you watch at home. The image is scanned and displayed very rapidly to provide the illusion of continuous movement.

FSTV operators are a tight-knit group because they are relatively few in number (compared to other segments of the hobby). Aside from the occasional band openings that send their signals bouncing over hundreds of miles, most FSTV conversations are local.

It's not uncommon for FSTV stations to take to the airwaves in the evening hours to share the news of the day. They may simply chat in *roundtable* format, each operator taking his or her turn before the camera.

"Look at this tomato," KE5ABC says as he holds the rotund vegetable before the lens. "I plucked it off the vine just after dinner. Isn't it a beauty?"

If you watch such a round-table, you may also see videos of a local club's Field Day operation, weather data or just about anything you can think of (as long as it's not commercial or obscene, of course!). Other FSTVers are more technically oriented. They meet on the air to compare software, antennas, cameras and so on.

Keep in mind that "tight-knit" does *not* necessarily mean "exclusive." FSTV operators are delighted to have new stations join the group. In fact, many TV enthusiasts spend a good deal of their time encouraging other hams to test the waters.

Avid FSTVers enjoy an extraordinary range of operating exploits. You can mount a simple camera, transmitter and associated electronics in a compact package and launch it as the payload beneath a helium balloon. Experimenters have amassed thrilling videotape recorded from the received signals of FSTV-equipped balloons at the edge of space, more than 100,000 feet up! The view from an altitude of 20+ miles is breathtaking.

Closer to the ground, hams combine hobbies by controlling their radio-controlled aircraft, boats and cars with a "pilot's-eye view" monitored from a camera mounted in their miniature craft. Amateur rocketeers mount FSTV equipment inside model rocket nosecones for a rapid ride up to 1000 feet or more, watching the world recede beneath their "eye in the sky."

Local public service agencies and government emergency preparedness offices are enthusiastic about having experienced hams provide television coverage of community events, drills and disasters. You can take your portable

Students at Southeastern Community College in Whiteville, North Carolina, prepare to launch a sophisticated FSTV-equipped rocket!

Radio-controlled airplane flying was never like this! With a tiny camera behind the nose gear and an FSTV transmitter in the fuselage, you can view the world as your aircraft sees it! (*photo by Ron Berkman, KA9CAP*)

FSTV station aboard a police or National Guard helicopter to survey the extent of forest fires, floods and storm damage. Parades, walkathons, bicycle races and other outdoor activities can be supported by skillful placement of mobile FSTV units sending video images to command posts. Trained weather observers can aid National Weather Service officials by letting them see developing storms or tornadoes firsthand.

Get a Glimpse of FSTV

Before you dive headfirst into FSTV, it's a good idea to check out the activity in your area. If you live in or near a large city, you have a decent chance of finding some FSTV stations on the air. In rural locations, however, FSTV operators may be hard to find.

Many operators rely on dedicated FSTV repeaters to extend their range. FSTV repeaters are similar to voice repeaters (see Chapter 2), except that they relay television signals. Like voice repeaters, FSTV repeaters are usually placed on hilltops, building roofs or other high-altitude locations. They also use a considerable amount of RF output power.

Because of their strong signals, FSTV repeaters are relatively easy to monitor. Get your hands on a recent copy of *The ARRL Repeater Directory* and you'll find a list of FSTV repeater frequencies.

If you think you have FSTV activity nearby, the next step is to watch it! Many portable LCD televisions can tune FSTV frequencies directly. The same is often true of cable-ready TVs and VCRs. To watch FSTV with your own receiver, all you have to do is tune to the appropriate "channel" (see **Table 6-1**).

If you don't mind spending about $100 to test the waters, you can purchase an FSTV *downconverter* from one of the vendors listed in the Info Guide. A downconverter takes 70-cm FSTV signals and converts them "down" to TV channel 3 or 4. All you need then is a conventional TV to watch the fun. A list of

Richard Logan, WB3EPX, enjoys operating FSTV from his home station.

Table 6-1

Cable TV Channels by Radio Frequency

Note: These are *cable* channels, not UHF TV channels.

Channel	MHz
57	421.25
58	427.25
59	433.25
60	439.25

Table 6-2

Popular 70-cm FSTV Frequencies (MHz)

Video	Audio	Use
421.25	425.75	Repeater outputs
426.25	430.75	Repeater inputs or outputs
434.00	438.50	Simplex and repeater inputs
439.25	443.75	Simplex, repeater inputs and outputs

commonly used 70-cm ATV frequencies appears in **Table 6-2**.

Unless you live in the shadow of a powerful FSTV station or repeater, you'll need a beam antenna to receive a watchable picture. Try a small, portable beam connected to your receiver through a short piece of low-loss coax. Point the beam out a nearby window and do your aiming by hand (this is just an experiment, after all).

The cheapest and easiest way to get your first glimpse of FSTV is through another ham's station. Start asking around at the next club meeting, or on your local repeaters. I'll wager that you'll find at least one ham who is more than happy to invite you over to see his or her station.

Planning Your FSTV Station

Unless you choose to set up the ultimate station, FSTV is very affordable (see **Figure 6-1**). If you happen to have a typical home video camera or camcorder and a basic TV set (black and white is fine), you can purchase an FSTV transceiver for about $400. Add perhaps $75 for a reasonable antenna and a few dollars for cables and miscellaneous odds and ends.

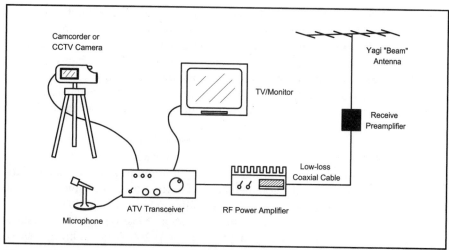

Figure 6-1—A diagram of a typical FSTV station.

Antennas, Amplifiers and Preamplifiers

Don't attempt permanent FSTV with anything other than a beam antenna. An FSTV signal is several *megahertz* wide. This means that your RF output is being spread over a broad chunk of spectrum. The more you spread RF power, the less punch it has on the receiving end. (That's why commercial TV stations often run several *megawatts* of effective radiated power!) If you expect reasonable range, you must focus your signal as much as possible with a beam antenna.

The good news is that a high-gain multielement Yagi for 434 MHz is only two or three feet long. Antennas for 902 and 1296 MHz are even smaller. Yes, you'll need a rotator to aim your antenna, but an inexpensive TV antenna rotator ($60 or less) is more than adequate. If you live in an apartment or condominium, a small beam and a rotator will fit nicely in almost any attic.

When choosing your FSTV antenna, don't forget about *polarization*. Most antennas used in FSTV applications are either horizontally or vertically polarized. Yagi antennas, for example, can be horizontal or vertical depending on how you mount them. Using the proper polarization is important. If you're vertically polarized and the station you're talking to is horizontally polarized, you'll both experience a noticeable loss in received signal strength.

So which antenna polarity should you use? It all depends on the FSTV activity in your area. If you intend to operate through an FSTV repeater, vertical polarization is best since most repeaters use vertically polarized antennas. For direct contacts—or FSTV DXing—horizontal polarization is the standard. Some antenna manufacturers allow you to enjoy the best of both worlds with dual-polarization antennas. These are basically two Yagi antennas mounted on the same boom. One is in the horizontal position and the other is vertical. An antenna switch (often remotely controlled) is used to select one antenna or the other.

If you've read the previous chapters, you can probably guess what's coming next. Yes, it's the low-loss-coax mantra! I've said it several times already, but it must be repeated again: Use the lowest-loss coax that you can afford. When it comes to FSTV, you're dealing with UHF signals. Loss can be horrendous at UHF if the SWR is even somewhat elevated.

In marginal areas, a receiver preamp mounted at the antenna can be a big help. (It's probably a good idea no matter where you live.) If you have a powerful TV broadcast transmitter nearby, you might have to add some high- or low-pass filters to keep the commercial station from overwhelming your receiver.

If you aren't within a dozen miles or so of a local FSTV repeater, think about adding a power amplifier to your station. Most typical FSTV transmitters put out 1 to 5 W, and boosting that to 50 or even 100 W could bring your signal up from barely copyable to sharp and clear!

Your FSTV Transceiver

Although it's the most expensive component of your station, it is also the easiest to set up and use. FSTV transceivers are designed for virtual plug-and-play operation. There are just a few controls and a couple of jacks to connect your camera, microphone and TV monitor. Check the Info Guide for a list of FSTV equipment vendors.

The Camera

If you have a camcorder in the house, you're in luck. You can use that camcorder as your FSTV station camera! Almost every camcorder on the market has a video-output jack for recording to VCRs and so on. You can use the same jack to bring the camera's signal to your FSTV transceiver.

If you can't afford a camcorder, look for a used *closed circuit* (CCTV) camera. Most of these cameras are black and white and their resolution (image detail) isn't the greatest, but they'll be sufficient to get you started. If you search the ham catalogs and the advertising pages of *QST* magazine, you'll often find tiny cameras retailing for $100 or less.

Used CCTV cameras are often found at hamfest flea markets, sometimes for as little as $30. If you buy a used CCTV camera, be prepared to do a little work. You may have to build a special power supply, or you may have to troubleshoot a few problems. Even so, if you purchase a camera that's in decent shape, you'll be on FSTV at a bargain price!

You're on the Air!

How do you call "CQ" on FSTV? If you want to be informal, just sit down in front of the camera and make yourself comfortable. Activate your transmitter and say, for example, "This is KDØXYZ in Hannibal, Missouri. Anyone around?" Switch back to the receive mode and wait for a response. If no one replies, choose another frequency (or repeater) and try again.

Another approach is to broadcast an image of your call sign for several seconds while announcing your invitation. Your call sign should be large enough to fill the entire screen. Bold, black letters on a white background will make it easier to read at greater distances. (Black lettering on white cardboard will do nicely.) Computers or character generators can also be used to create attractive ID screens. Let your imagination be your guide!

You can establish contact with another station on a single frequency (simplex), or use an FSTV repeater with separate input and output frequencies (duplex). The same operating procedures apply in either case. Some FSTV repeaters transmit *beacons* (usually composed of its call sign and some graphics) to help you find them. Other FSTV repeaters transmit their identifications at regular intervals (on the hour and half-hour, for example). When you transmit, the beacon will disappear and your video signal will be relayed through the system. When you stop transmitting, the beacon will return.

When you finally establish contact, the first order of business is to exchange signal reports so that each of you will know how well your signals are being received. In FSTV we use the "P" system to describe the amount of noise (or "snow") in the picture. A P5 picture, for example, is excellent with no visible noise. A P1 picture is very weak with a great deal of noise (see **Figure 6-2**).

After that, anything goes! You can talk about whatever comes to mind. Unlike other Amateur Radio modes, the person at the other end will be able to see your gestures and facial expressions as you speak. This gives FSTV a unique personal dimension that is difficult to achieve on SSB, CW, FM or packet.

The 2-Meter Connection

In many areas of the country, it's a popular practice to use a 2-meter FM simplex frequency—or a repeater—for local FSTV coordination. This allows FSTVers to make critical adjustments *while transmitting*. Since mutual interference between the 2-meter and 420-MHz equipment is minimal, FSTVers can get instant comments on their signal quality and make additional adjustments as necessary.

"AA8QVC from N8SVN. Your signal is getting worse. You turned your beam too far."

"There. The antenna should be pointing directly at the repeater. How's that?"

"Much better, Tom. I can see you and the chair you're sitting in, but everything else is pretty dark."

"That's a lighting problem. Let me switch on my desk lamp."

"Nice! Now I can see everything. Say, that's an impressive movie poster you have on your wall. Where did you get it?"

Many enthusiasts will call CQ on 2-meter FM and 70-cm FSTV simultaneously. This technique is effective in attracting the attention of local operators

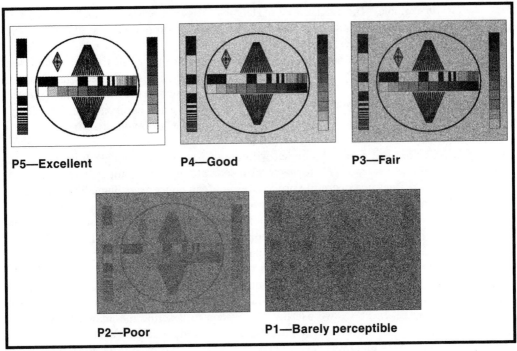

P5—Excellent   **P4—Good**   **P3—Fair**

P2—Poor   **P1—Barely perceptible**

Figure 6-2—The FSTV picture quality-reporting system. A P5 picture is very clear and easy to see. On the other hand, a P1 image cannot be seen at all.
P1—Barely perceptible
P2—Poor
P3—Fair
P4—Good
P5—Excellent

who may be monitoring 2 meters while their FSTV equipment is inactive. If FSTV coordination is taking place on 2 meters in your area, try to find the frequency. (144.34 MHz is popular.) By listening to the chatter you'll get a pretty good idea of how many FSTV stations are on the air, where they are located and when they are active.

FSTV DXing

There is more to FSTV than chatting with your local friends. You'll also have opportunities to make contacts with FSTVers hundreds of miles away. It doesn't happen every day, but when it does, FSTV DXing offers genuine excitement!

If you plan to work FSTV DX via direct, single-frequency contacts, you'll need to invest in the best equipment you can afford. Antenna-mounted preamplifiers are mandatory along with RF power amplifiers and low-loss coaxial cable or Hardline. Some FSTV DXers use a single, high-gain Yagi antenna (horizontally polarized). Others

The W5KPZ FSTV repeater in Tyler, Texas. Repeaters like these provide wide coverage for even small, low-power FSTV stations.

My own SSTV "CQ" as received by a friend 1000 miles away.

When it comes to video DX, John, KD0LO, is an expert! He has enjoyed FSTV contacts with stations more than 400 miles from his home in St Louis, Missouri. (*photo by Bill Brown, WB8ELK*)

use several Yagis working together in a carefully designed assembly fed by the primary coaxial line. This is often called an *array*. Depending on its construction, an array is capable of very high performance in DX applications.

You can also work DX through your local FSTV repeater. When the band is open, many FSTVers will attempt to reach repeaters in distant locations. Don't be surprised if you call CQ on your repeater one evening and receive a reply from an FSTV DXer!

The best times for FSTV DX are around sunrise and right after sunset. If your family TV is connected to an outside antenna, tune through the commercial UHF channels. If you begin to see distant stations, especially on channels 14 through 30, there may be a band opening in progress!

S-L-O-W Scan TV (SSTV)

Throughout this chapter we've been talking about *fast-scan* TV—the real-time, full-motion television you're accustomed to seeing. But hams also exchange still images on the HF bands through a technique known as *slow-scan TV* or *SSTV*.

Don't misunderstand—fast-scan TV is a fun mode, but the transmitted range is limited to a few hundred miles under the best conditions because FCC Rules limit the use of FSTV to frequencies above 420 MHz. In addition, special transmitters and receivers are required.

But what if you wanted to share images over *thousands* of miles? You could take a fixed (nonmoving) image and slowly scan it line by line, converting the color and brightness variations into audio tones. Feed the audio tones to an ordinary SSB voice transceiver and you can send this information almost anywhere in the world. On the receiving end, the audio tones are translated back into an image on a computer screen. That's SSTV!

Thanks to the proliferation of personal computers and sound cards, it has never been easier to enter the world of SSTV. Most SSTV activity takes place on 20 meters, but you'll find it on other bands as well (28.680 MHz is a popular hangout for 10-meter SSTV). Hams licensed as Technicians should note that SSTV can be used on VHF, too. You'll find activity on 6 meters from time to time, and a few amateurs have even swapped SSTV images via satellite!

As Close as Your Computer

Most of the software/hardware packages for SSTV are designed around IBM-compatible PCs. You don't need an ultra-fast, sophisticated machine; a Pentium-class PC should do just fine. There are a number of SSTV programs available; some are freeware, some are shareware, and others are strictly commercial. Most of these programs run under the various flavors of *Windows*.

SSTV on a PC is more than just software, though. We have to translate analog SSTV tones into digital data the computer can understand. When it's time to transmit, the same process must occur in reverse. So we need an interface to bridge the gap between the digital and analog worlds. Sound cards are perfectly suited for

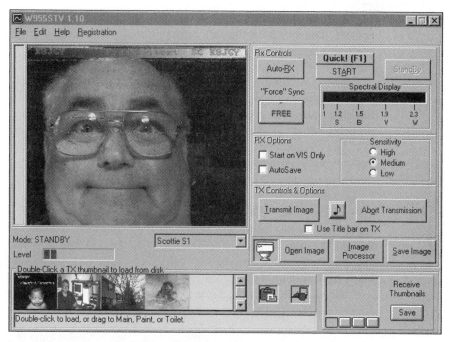

The smiling face of K8JGY on SSTV. The software is _W95SSTV_, a sound-card based SSTV package for _Windows_.

this task—analog-to-digital conversion (and vice versa) is what they do best. That's why you'll find SSTV software written for PCs equipped with SoundBlaster (or compatible) cards. The wiring to and from your transceiver is identical to the setup needed for the HF digital modes (see Chapter 3).

How Do We Make an SSTV Picture?

Receiving an SSTV image is easy and fun, but eventually you'll want to start sending images of your own. The first step is to create a picture. There are many ways to do this.

Digital cameras are becoming quite affordable (there are several on the market right now that cost less than $200). You can snap a picture, dump the image data to your PC, and send the data over the air. If you don't want to buy a digital camera, how about a hand-held or flatbed color scanner? Many are available for under $100. Pop a picture into the scanner and within seconds you have an image you can transmit.

Do you own a camcorder or a VCR? Video digitizers such as Snappy, Computer Eyes, Ventek, Video Logic Captivator and others can freeze and capture video images (live or prerecorded). Just save them to your hard drive and transmit!

If you're running Microsoft *Windows*, chances are you have *Paintbrush* on your PC. Did you know that you can use *Paintbrush* to create artwork that you can transmit via SSTV? Of course, if you own more sophisticated software such as *Paint Shop Pro*, you can really get cooking!

By the way, most SSTV programs have a text function that lets you place "type" on the image to add information, comments or simply your call sign. These functions, however, are often limited in choices of fonts, colors and so on. You're probably better off using *Paintbrush* or other software to dress up your images.

And what kind of images can you send? Anything is fair game, as long as it isn't offensive or in poor taste. Show the world your rig, antenna, your home, your family or...yourself.

On-the-Air Operating Suggestions

I suggest that you begin your SSTV adventure by simply listening. Take a few minutes this weekend and eavesdrop on the SSTV groups. What you'll hear are brief announcements such as, "Here's a picture of my new antenna. *Scottie One!*" followed by the *deedle-deedle* sound of the SSTV data signal.

"Scottie one" isn't the name of his antenna, it's the mode he is using to send

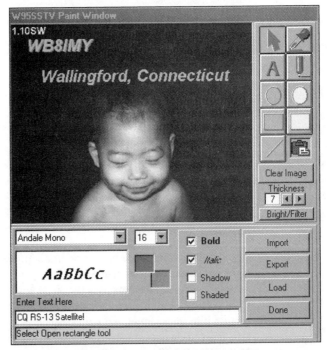

Many SSTV programs include an image editor where you can add text and other effects to existing images.

the image. There are 27 different modes in use, with the most popular being Scottie One and Martin One. Everyone who is receiving your picture must have his or her software set to the proper mode. That's why you should announce the mode before transmitting.

The entire transmission can take as long as a minute or two. After that you'll hear the other operators jump in with quick comments such as, "Nice-looking antenna, Joe!" If a distant station failed to copy the image (noise and interference can scramble an SSTV picture), another station might act as a relay by retransmitting the data.

After you've listened for a while, fire up your SSTV receive system and copy a few images. The quality and beauty will amaze you! You'll want to save some of these to disk and build a library of your own.

When you think you're ready to send some SSTV of your own, join the group and give it a try. You'll find that SSTVers are a pleasant bunch and they always welcome new operators.

Expand Your Vision

One of the most fascinating aspects of SSTV is that you get to *see* the people and places you've only *heard* in the past. There's something about seeing the face of a friend that adds a new dimension to any conversation. And just imagine what you'll see when the operator points his camera out the window to give you a minutes-old glimpse of his neighborhood. (If the nearby tree branches are swinging in the breeze, you'll see some odd "tears" in the image, but that's okay!) Now I ask you, is this slick or *what*?

7 | A Conversation with the World

In Chapter 3 we discussed the pleasures of digital communication on the HF bands—the amateur frequencies below 30 MHz. In Chapter 6 we described how you can use slow-scan television to send images around the world on the same frequencies. It's time to step back and take a good look at that strange portion of the radio spectrum where the fates of signals are governed by the Sun and a high-altitude layer of our atmosphere known as the ionosphere. A signal transmitted on these frequencies can go almost anywhere on the Earth.

NINE FLAVORS OF ICE . . . ER, *SPECTRUM*

There are nine groups of frequencies—or *bands*—in the HF/MF region where hams are allowed to operate. Each has its own "flavor." Let's take a brief look . . .

160 Meters: 1.8 to 2 MHz

This is the "basement" of Amateur Radio, the lowest band that we can use. It's almost *un*usable during daylight hours, but nighttime signals on 160 can travel hundreds or even thousands of miles. Noise levels are high, especially during the summer months when thunderstorms march across the landscape. This makes 160 meters primarily a "winter" band. You'll find a mix of CW and SSB activity on this band, although most of the long-distance work (DX) takes place on CW. Because of the large antennas required for 160 operating, it isn't a band for everyone. As a result, crowding is at a minimum.

80 Meters: 3.5 to 4 MHz

Like 160 meters, 80 meters is considered a nighttime band. Even so, 80 is also good for daytime communication out to a few hundred miles.

Depending on atmospheric conditions, 80 meters can offer worldwide communication at night. Even under mediocre conditions, contacts between the East Coast of the US, for example, and Europe are common. Like 160, this band also suffers from high noise levels, so it is definitely at its best during the winter.

The voice portion of 80 meters is inhabited primarily by SSB (single sideband) voice operators, although you'll hear a couple of AM signals here and there. When the band is in decent shape, the voice portion can become extremely crowded. This is particularly true between 3.85 and 4 MHz.

You may also find wall-to-wall signals in the CW segment. Chasing international contacts on 80-meter CW is a favorite pastime.

40 Meters: 7 to 7.3 MHz

Forty meters is a transition band. It shares characteristics with the lower and higher HF bands.

During the day, 40 is excellent for communication over distances of about 500 miles or so. Phone operators enjoy meeting on 40 for late-morning chats. *Nets*—groups of hams who congregate on the air for a specific purpose—also exploit the advantages of 40 meters in the daylight hours.

At night, 40 meters opens to the world. CW operators enjoy global range in the lower portion of the band. SSB enthusiasts would relish the same contacts were it not for severe interference from shortwave broadcast stations. These high-powered RF blasters decimate the 40-meter voice segment throughout much of the world at night.

30 Meters: 10.1 to 10.150 MHz

This is strictly a CW and digital band. No voice operating is allowed. In addition, you can't use more than 200 W output.

Thirty meters is good for DX work during the daylight hours, and up to several hours after dark. The CW operators gather in the lower portion of the band to chase international contacts. Thirty meters is also a terrific band for low-power (QRP) CW activity. The digital folks occupy the upper portion.

20 Meters: 14 to 14.350 MHz

Twenty meters is the queen of the DX bands. It's open to just about every corner of the world at various times of the day. During the peaks of the 11-year solar cycle, 20 meters will remain open throughout the night. Otherwise, it tends to shut down after local sunset.

All modes are hot and heavy on 20 meters! SSB reigns supreme in the upper half of the band. You may have a tough time finding a clear frequency between 14.225 and 14.350 MHz, especially on weekends. CW occupies much of the bottom portion. DX chasers often hunt between 14 and 14.050 MHz. Low-power (QRP) CW operators hang out around 14.060 MHz. The digital modes can be heard anywhere from 14.065 to 14.120 MHz.

17 Meters: 18.068 to 18.168 MHz

This daytime band also offers worldwide communication, though it is never crowded. SSB seems to be the dominant mode of communication on 17 meters, although you'll find a few CW stations and the occasional digital operator.

Seventeen meters is best during the peak years of the solar cycle. Even so, it can often provide global DX even in the worst years.

15 Meters: 21 to 21.450 MHz

This is a hot DX band when the solar cycle reaches its peak. In cycle doldrums, 15 meters can provide some mediocre DX, but it's usually limited to sporadic openings over 500 to 1,000 miles.

Like 17 meters, 15 is a daytime band. In the best years, it opens in the late morning and closes a few hours after dark. You'll find a mix of SSB and CW on this band, although the SSB operators usually predominate. The digital segment can also be quite active.

12 Meters: 24.890 to 24.990 MHz

When the solar cycle is up, so is this band—at least in the daylight hours. When the cycle is down, 12 meters is a wasteland. Even in the best years, contacts are few and far between. This has more to do with a lack of interest than anything else. You'll hear the odd SSB and CW conversation, but most operators prefer to migrate to the next-highest band—10 meters.

10 Meters: 28 to 29.700 MHz

During the best years of the solar cycle, 10 meters is one of the hottest DX bands around. At solar-cycle peaks, the ionosphere absorbs relatively little of your signal at this frequency. It simply bends it back to Earth thousands of miles away. As a result, even low-power stations can use 10 meters to work the world with ease.

When sunspots are scarce, so are contacts on 10 meters. In fact, many hams consider 10 meters to be worthless during the *solar minimum*. That's an exaggeration. While it's true that you won't make too many DX contacts during the low points in the cycle, the band frequently opens for conversations over hundreds of miles.

When it's open, 10 meters is usually a daytime band. It opens in the late morning and shuts down at dark. SSB activity is most heavily concentrated in the segment from 28.300 to 28.500 MHz. CW is relatively rare and so are the digital modes. At the top end of the band you'll find a segment from 29.500 to 29.700 that's dedicated to FM operating.

SETTING UP YOUR HF STATION

Grab a piece of paper and prepare for a pop quiz. (Or you can write in this book if you dare deface such a valuable piece of literature!)

1.   I want my first HF station to operate . . . (check one)
- ☐ On one band
- ☐ On several bands

2.   I want to operate . . . (check one)
- ☐ SSB
- ☐ CW
- ☐ SSB and CW
- ☐ Digital
- ☐ All of the above

3.   I don't want to spend more than: _____

Congratulations! You've just established three major criteria for building your first HF station. Mark this page because you'll refer to it often as we get down to brass tacks. Right now, in fact . . .

Questions and Answers About Lightning Protection

By Mike Tracy, KC1SX

Q: I haven't had any lightning problems yet. Why do I need protection?

A: When most hams think of lightning protection, they immediately think about ways to protect their station equipment. Although that is certainly important, you should have far more concern for the health and welfare of yourself and your family. Each year, lightning is responsible for the deaths of over 400 people in the US. Several hundred more suffer from injuries caused by lightning, such as burns, shock and other damage to the body's more vulnerable parts.

Q: How much of a threat do I face?

A: The number of local thunderstorm days per year in this country ranges from 1 to 100, depending on where you live. If you live in a location with a single thunderstorm day, that means that you have at least one opportunity for disaster to strike. The total number of strikes per year is more than 40 million. However impressive these statistics may seem, keep in mind that they do not include all lightning strikes. Lightning can occur even without a thunderstorm—whenever and wherever there is a sufficient charge build-up.

Many things are involved in determining the likelihood of a strike at your home. A brief list includes the type of structure, the materials it's made of, the location relative to other structures and so on.

Other reasons for lightning protection include fire prevention and protection of sensitive electronic equipment. Property damage statistics indicate that lightning causes over 40 million dollars damage annually to buildings and equipment in the US.

In addition, your equipment can also be damaged by other electrical disturbances such as power line switching transients and voltage surges, as well as static build-up on outside wires and antennas.

Q: But I already have lightning protection. My station is grounded and I added a lightning arrestor to the coax.

A: Your situation is typical of many hams: a single copper rod driven into the earth as a

The Radio

This is the category where sticker shock may hit you the hardest—depending on your answers to questions 1 and 2. New multiband HF transceivers are not cheap. You'll pay at least $600 for a 100-W transceiver that can do SSB, CW and the digital modes. If you insist on a radio that has every conceivable goodie you can imagine, you could fork over as much as $4000!

If you have the money lying around, go for the gusto. Today's amateur transceivers are excellent bargains when you consider the level of technology they represent. The radio you buy today will probably still be plugging away decades from now. To make the best purchase decision, read the Product Reviews that appear each month in *QST* magazine. We've also compiled past reviews in our *Radio Buyer's Sourcebooks*.

But if a $600 price tag jolts you into cardiac arrest, look into *used* equipment. Many amateur dealers who advertise in *QST* stock used gear and even offer war-

---

station equipment ground and an in-line coax lightning arrestor, often mounted in the shack at the operating position. For lightning protection this sort of installation is not adequate. It may even be an invitation to disaster.

Q: What steps should I take first to add protection to my shack?

A: The most important thing to do is to keep lightning outside of your home. This includes disconnecting your equipment from the feed lines and power sources, providing a proper station ground and adding protective devices to your installation.

As *The ARRL Antenna Book* states, "The best protection from lightning is to disconnect all antennas from equipment and disconnect all equipment from power lines." When lightning strikes, it will always try to find the shortest electrical path to ground. Unless you disconnect your station equipment, you're giving the strike a good return path through your equipment!

The easiest way to remember to do this is to disconnect your station whenever you're not using it. To prevent lightning from using your feed lines as a sneak path into your shack, disconnect them outside. If you disconnect your coax and leave it lying on the floor, lightning can jump a gap of several feet to your grounded equipment. Remember that it has already traveled quite a distance through the air. A few more feet of atmosphere won't stop it (this phenomenon is known as a "side flash").

The slick approach is to install an entrance panel for your feed lines and control cables. Place the panel ground connection on the outside of your home. Don't attach it to an inside source such as the power company ground or a cold water pipe. This panel will provide a convenient disconnect point for your equipment, as well as a place to mount feed line and control cable transient protectors.

Q: I can do that. But what about my station ground system?

A: Proper grounding is critical to lightning protection. Lightning contains energy in a wide range of frequencies (which is why you can hear "static crashes" on an AM radio when a storm approaches). You must provide a low-impedance path to ground for the energy. A single ground rod will not suffice as a lightning ground. The basic idea is to give the strike energy a place to dissipate.

ranties of 30 to 90 days. You'll pay a little more, but the warranty is worthwhile when you're talking about "preowned" radios. The alternative is buying directly from another ham. Hams are a pretty trustworthy group, but there are some who suffer from HDD—Honesty Deficit Disorder. If you shop the hamfest flea markets, classified ads, Web auction sites, do so with your eyes open. Whenever possible, try before you buy.

Shun any used transceiver that contains vacuum tubes. Tubes are quickly becoming ancient technology, and replacements are getting harder to find. As a rule of thumb, stick with solid-state gear that's no more than 10 years old. With a little savvy shopping, you'll be able to locate a good used radio with a price tag of $400 or less. And never buy a rig that doesn't include an operator's manual. It will help you avoid hours of frustration.

If $400 still has you reaching for the smelling salts, it's time to consider a *single*-band, *single*-mode transceiver. That almost always means a low-power QRP transceiver. A pre-built transceiver will set you back about $180. If you can put together an electronic kit, expect to spend between $100 and $150.

You can have a lot of fun with just a few watts of power on CW. Under the right conditions, 5 W will go just as far as 100 W. Your signal may not pin the other fellow's S meter to the wall, but it really doesn't matter. If he can understand your transmission, that's all that counts. You'll reap the benefit of using a radio that doesn't cost much, doesn't require much space, and doesn't interfere with your neighbor's TV or other electronic toys.

Amplifiers

If you feel that the RF output of your radio is too anemic, will an amplifier really help? Hmmm . . . maybe. Boosting your output from 100 W (the typical output of a multimode, multiband radio) to 1500 W *will* have a positive effect at the receiving end. But it may also have a negative effect on every electronic device in your home, your neighbor's home, *his* neighbor's home . . . you get the idea.

Amplifiers also make substantial impacts on bank accounts. A 1500-W top-of-the-line powerhouse can set you back as much as $3000. Of course, you may also need some electrical work in your home to route a high-current ac line (usually 220 V) to your radio room. That may come at a price, too.

Am I advocating that you "say no" to HF amplifiers? Not at all. Amps have their applications, and if they didn't work they wouldn't sell as well as they do. When band conditions are mediocre or poor, an amplifier might make the difference between being heard or being just another whisper in the static. They're the favorite tools of competitive contesters and DXers who *must* be heard, no matter what.

The problem with HF amplifiers is their high cost and their tendency to cause interference. A 100-W radio by itself is capable of causing plenty of interference to televisions, telephones, stereos, VCRs, alarm systems and so on. Boost that power by a factor of 15 and guess what happens? If you don't have neighbors nearby, or if you feel competent to track down and cure the interference problems, perhaps an amp is in your future (assuming you have the cash, too).

Antennas

When you think of an antenna for the HF bands, what sort of image appears in your mind? Do you see a gleaming steel tower, majestically supporting a huge rotating beam antenna? That's the vision most of us conjure. The radio tower is an ancient icon in our hobby. It symbolizes the art and mystery of radio itself.

But ham antennas come in almost every design imaginable. Some are little more than strands of finely tuned wire. Others are thorny javelins of polished metal. Some sit atop towers. Others don't. Some look like monstrous spider webs, while others are modern-art sculptures of aluminum tubing.

There are far too many antenna designs to discuss in the few pages we have available. If you want the complete story, pick up the latest edition of *The ARRL Antenna Book*. This is the ultimate A-to-Z antenna guide, and it even includes software.

In the meantime, let's take a look at several of the most popular antennas you're likely to encounter. We'll begin with the big kids on the block . . . the beams.

HF Beams

When hams speak of beam antennas, they usually mean the venerable Yagi and quad designs (see **Figure 7-1**). These antennas focus your signal in a particular

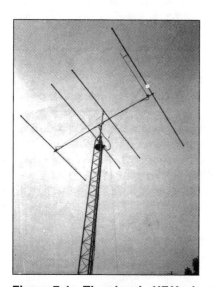

Figure 7-1—The classic HF Yagi antenna (left) is often found atop tall towers. Another tower hog is the *quad* antenna (right). This three-band HF quad is an impressive sight above any home! Both antennas require heavy-duty rotators so that they can be aimed in the desired directions.

direction (like a flashlight). Not only do they concentrate your transmitted signal, they allow you to focus your *receive* pattern as well. For example, if your beam is aimed west you won't hear many signals from the east (off the "back" of the beam).

The problems with HF beam antenna systems are size and cost. HF beams for the lower bands are *big* antennas. At about 43 feet in width, the longest element of a 40-meter coil-loaded Yagi is wider than the wingspan of a Piper Cherokee airplane. Even a 10-meter beam is about 18 feet across.

And a multiband (20, 15 and 10 meter) beam antenna and a 75-foot crank-up tower will set you back *at least* $2500. Then add about $500 for the antenna rotator (a beam isn't much good if you can't turn it), cables, contractor fees (to plant the tower in the ground) and so on. In the end, you'll rack up about $3000.

If you have that much cash burning a hole in your pocket, by all means throw it at a beam antenna and tower. The rewards will be tremendous. Between the signal-concentrating ability of the beam and the height advantage of the tower, you'll have the world at your fingertips. Even a beam antenna mounted on a roof tripod can make your signal an RF juggernaut. If you intend to become an avid DX chaser or contester, toss another $2000 into an amplifier and you'll be slugging it out with the best of them.

In truth, only a minority of hams can afford towers these days. Those who manage to scrape together the necessary funds occasionally find themselves the targets of angry neighbors and hostile town zoning boards. (They don't appreciate the beauty of aluminum and steel like we do!)

But do you *need* a beam and a tower to enjoy Amateur Radio? This issue isn't whether they're worthwhile (they are). The question is: Are they absolutely necessary? The answer, thankfully for most us, is *no*.

Single-Band Dipoles

You can enjoy Amateur Radio on the HF bands with nothing more than a copper wire strung between two trees. This is the classic *dipole* antenna. It comes in several varieties, but they all function in essentially the same way.

Single-band dipoles are among the easiest antennas to build. All you need is some stranded, noninsulated copper wire and three plastic or ceramic insulators (see **Figure 7-2**). A ¹/₂-wavelength dipole is made up of two pieces of wire, each ¹/₄-wavelength long.

Calculating the lengths of the ¹/₄-wavelength wires is simple. Just grab a calculator and perform the following bit of division:

Length (feet) = 468/frequency (MHz)

Actually, you should add about six inches to the results of your calculations. You'll need that length margin to trim for the lowest SWR.

Join the two wires in the center with an insulator, then place insulators at both ends. Solder the center conductor of your coaxial cable to one side of the center insulator. (It doesn't matter which side.) Solder the shield braid of your

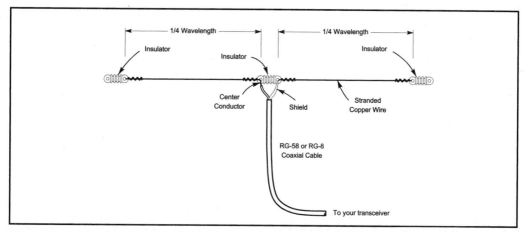

Figure 7-2—The ¹/₂-wavelength dipole is one of the easiest HF antennas to build. All you need is stranded copper wire and three insulators. You can feed a dipole with RG-58 or RG-8 coax, but you must trim the length of the antenna to achieve the lowest SWR.

cable to the other side. Connect ropes, nylon string or whatever to the end insulators and haul your antenna skyward. Get it as high as you can and as straight as possible. Don't hesitate to bend your dipole if that's what it takes to make it fit.

Once your dipole is safely airborne, fire up your transmitter and check the SWR at many points throughout the band. (It helps if you can plot the results on graph paper.) If you see that the SWR is getting *lower* as you move lower in frequency, your antenna is too long. Trim a couple of inches from each end and try again. On the other hand, if you see that the SWR is getting *higher* as you go lower in frequency, your antenna is too short. You'll need to *add* wire to both ends and make another series of measurements.

When you've finished trimming your dipole, you'll probably end up with an SWR of 1.5:1 or less at the center frequency, rising to 2:1 or somewhat higher at either end of the band. Don't expect a 1:1 SWR across the entire band.

Trap Dipoles and Parallel Dipoles

For multiband applications, you'll often find the *trap* dipole (**Figure 7-3**) and the *parallel* dipole (**Figure 7-4**). Traps are tuned circuits that act somewhat like automatically switched inductors or capacitors, adding or subtracting from the length of the antenna according to the frequency of your signal. The parallel dipole uses a different approach. In the parallel design, several dipoles are joined together in the center and fed with the same cable. The dipole that radiates the RF is the one that presents an impedance that most closely matches the cable (50 Ω). That matching impedance will change according to the frequency of the signal. One dipole will

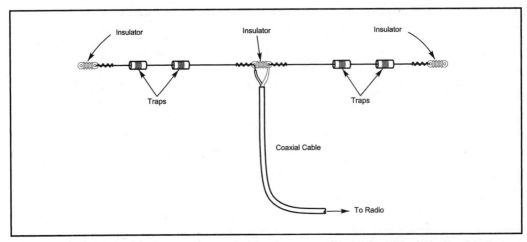

Figure 7-3—A trap dipole uses tuned circuits known as "traps" to electrically shorten or lengthen the antenna. Some elaborate trap dipoles offer coverage on many bands. You can design and build your own trap dipole, if you have enough experience. Otherwise, you're better off buying one premade.

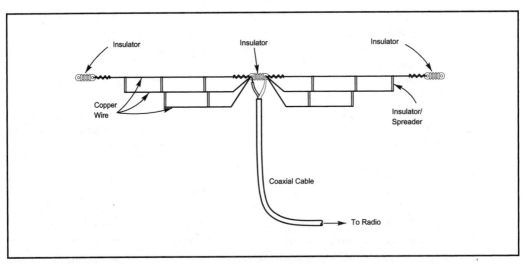

Figure 7-4—You can connect several different dipoles at the center and create the *parallel dipole* antenna. The dipoles tend to interact with each other, so building a parallel dipole usually entails *lots* of trimming and measuring. If you don't think you're up to the task, several *QST* advertisers offer prebuilt parallel dipole products.

offer a 50-Ω match on, say, 40 meters, while another provides the best match on 20 meters.

Obviously, these designs are somewhat more complicated than monoband dipoles, although many hams *do* choose to build their own. (See *The ARRL Antenna Book* for construction details.) If you don't have time or desire to tackle a trap or parallel dipole, you'll discover that many *QST* advertisers sell prebuilt models.

Random-Length Dipoles

You can also enjoy multiband performance *without* traps, coils, fans or other schemes. Simply cut two equal lengths of stranded copper wire. These are going to be the two halves of your dipole antenna. Don't worry about the total length of the antenna. Just make it as long as possible. You won't be trimming or adding wire to this dipole.

Feed the dipole in the center with 450-Ω *ladder line* (available from most ham dealers), and buy an *antenna tuner* with a *balanced output* (see **Figure 7-5**). Feed the ladder line into your house, taking care to keep it from coming in contact with metal, and connect it to your tuner. Use regular coaxial cable between the antenna tuner and your radio.

You can make this antenna yourself, or buy it premade if you're short on time. A 130-foot dipole of this type should be usable on almost every HF band. Shorter versions will also work, but you may not be able to load them on every band.

Ladder line offers extremely low RF loss on HF frequencies, even when the SWR

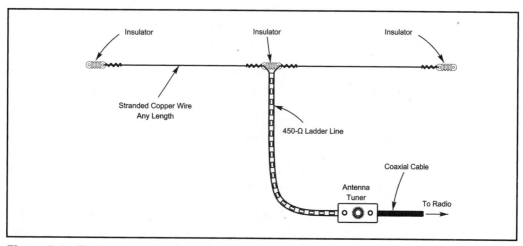

Figure 7-5—The random length dipole is simplicity itself! Just put up as much antenna wire as possible, feed it in the center with 450-Ω ladder line, and you're on the air on several bands.

is relatively high. Just apply a signal at a low power level to the tuner and adjust the tuner controls until you achieve the lowest SWR reading. (Anything below 2:1 is fine.) You'll probably find that you need to readjust the tuner when you change frequencies. (You'll *definitely* need to readjust it when you change bands.)

You may discover that you cannot achieve an acceptable SWR on some bands, no matter how much you adjust the tuner. Even so, this antenna is almost guaranteed to work well on several bands, despite the need to retune.

So why doesn't everyone use the ladder line approach? The reason has much to do with convenience. Ladder line isn't as easy to install as coax. As I've already noted, you must keep it clear of large pieces of metal (a few inches at least). Unlike coax, you can't bend and shape ladder line to accommodate your installation. And ladder line doesn't tolerate repeated flexing as well as coaxial cable. After a year or two of playing tug o' war with the wind, ladder line will often break.

Besides, many hams don't relish the idea of fiddling with an antenna tuner every time they change bands or frequencies. They enjoy the luxury of turning on the radio and jumping right on the air—without squinting at an antenna tuner's SWR meter and twisting several knobs.

Even with all the hassles, you can't beat a ladder-line fed dipole when it comes to sheer lack of complexity. Wire antennas fed with coaxial cable must be carefully trimmed to render the lowest SWR on each operating band. (Go back and read Chapter 2 if you've forgotten what happens to RF when the SWR gets too high.) With a ladder line dipole, no pruning is necessary. You don't even care how long it is. Simply throw it up in the air and let the tuner worry about providing a low SWR for the transceiver.

Whichever dipole you finally choose, install it as high as possible. If a horizontal dipole is too close to the ground, the lion's share of your signal will be going skyward at a steep angle. Without wading chest deep into propagation analysis, the bottom line is that a high radiation angle is generally not good for long-distance communication. Forty to 70 feet is generally considered the ideal height range, but don't lose sleep if you fall short. Raise the antenna as high as you can and change the subject when you're asked about it. You'll still make lots of contacts.

Verticals

The vertical is a popular antenna among hams who lack the space for a beam or dipole. In an electrical sense, a vertical is a dipole with half of its length buried in the ground or "mirrored" in its counterpoise system. Verticals are commonly installed at ground level, although you can also place a vertical on the roof of a building.

At first glance, a vertical looks like little more than a metal pole jutting skyward. A single-band vertical may be exactly that! However, if you look closer you'll find a network of wires snaking away in all directions from the base of the antenna. In many instances, the wires are buried a few inches beneath the soil.

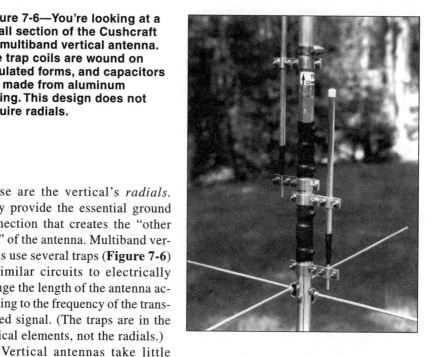

Figure 7-6—You're looking at a small section of the Cushcraft R7 multiband vertical antenna. The trap coils are wound on insulated forms, and capacitors are made from aluminum tubing. This design does not require radials.

These are the vertical's *radials.* They provide the essential ground connection that creates the "other half" of the antenna. Multiband verticals use several traps (**Figure 7-6**) or similar circuits to electrically change the length of the antenna according to the frequency of the transmitted signal. (The traps are in the vertical elements, not the radials.)

Vertical antennas take little horizontal space, but they can be quite tall. Most are at least ¼-wavelength long at the lowest frequency. To put this in perspective, an 80-meter full-sized vertical can be over 60 feet tall! Then there is the space required by all those radial wires. You don't have to run the radials in straight lines (see **Figure 7-7**). In fact, you don't even have to run them underground. But you *do* need to install as many radials as possible for each band on which the antenna operates. Depending on the type of soil in your area, you may get away with a dozen radials, or you may have to install as many as 100.

Contemplate spending several days on your hands and knees pushing radial wires beneath the sod. It isn't a pretty picture, is it? That's why several antenna manufacturers developed verticals that do not use radials at all. There are questions concerning the efficiency of these antennas, but many hams swear by them (as opposed to *at* them).

So how does the vertical stack up against the dipole when it comes to performance? If you have a generous radial system, the vertical can do at least as well as a dipole in many circumstances. Some claim that the vertical has a special advantage for DXing because it sends the RF away at a low angle to the horizon. Low radiation angles often mean longer paths as the signal bends through the ionosphere.

Without a decent radial system, however, the vertical is a poor cousin to the dipole. The old joke, "A vertical radiates equally poorly in all directions," often applies when the ground connection is lacking, such as when the soil conductivity

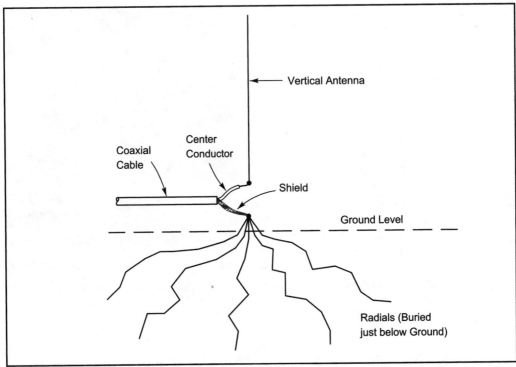

Figure 7-7—You don't have to lay down your vertical's radial wires in straight lines. Bend and twist them as much as necessary to fit your available space.

is poor. If you can't lay down a spiderweb of radials, dipoles are often better choices.

Random Wires

A random wire is exactly that—a piece of wire that's as long as you can possibly make it. One end of the wire attaches to a tree, pole or other support, preferably at a high point. The other end connects to the random-wire connector on a suitable antenna tuner (**Figure 7-8**). You apply a little RF and adjust the antenna tuner to achieve the lowest SWR. That's about all there is to it.

Random-wire antennas seem incredibly simple, don't they? The only catch is that your antenna tuner may not be able to find a match on every band. The shorter the wire, the fewer bands you'll be able to use. And did you notice that the random wire connects directly to your antenna tuner? That's right. You're bringing the radiating portion of the antenna right into the room with you. If you're running in the neighborhood of 100 W, you could find that your surroundings have become rather hot—*RF* hot, that is! We're talking about painful "bites" from the metallic

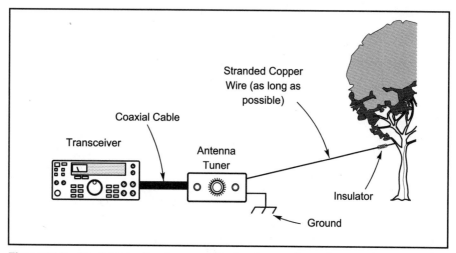

Figure 7-8—If a beam, dipole or vertical antenna doesn't work for you, consider the random end-fed wire.

portions of your radio, perhaps even a burning sensation when you come in contact with the rig or anything attached to it.

Random wires are fine for low-power operating, however, especially in situations where you can't set up a vertical, dipole or other outside antenna. And you may be able to get away with higher power levels if your antenna tuner is connected to a good Earth ground. (A random-wire antenna needs a good ground regardless of how much power you're running.) If your radio room is in the basement or on the first floor, you may be able to use a cold water pipe or utility ground. On higher floors you'll need a *counterpoise*.

A counterpoise is simply a long, insulated wire that attaches to the ground connection on your antenna tuner. The best counterpoise is ¼-wavelength at the lowest frequency you intend to use. That's a lot of wire at, say, 3.5 MHz, but you can loop the wire around the room and hide it from view. The counterpoise acts as the other "terminal" of your antenna system, effectively balancing it from an electrical standpoint.

Indoor Antennas

So you say that you can't put up an outdoor antenna of any kind? There's hope for you yet. Antennas generally perform best when they're out in the clear, but there is no law that says you can't use an outdoor antenna *indoors*.

If you have some sort of attic in your home, apartment or condo, you're in luck. Attics are great locations for indoor antennas. For example, you can install a wire dipole in almost any attic space. Don't worry if you lack the room to run the dipole in a straight line. Bend the wires as much as necessary to make the dipole fit into the available space (see **Figure 7-9**).

Of course, this unorthodox installation will probably require you to spend some time trimming and tweaking the length of the antenna to achieve the lowest SWR (anything below 2:1 is fine). Not only will the antenna behave oddly because of the folding, it will probably interact with nearby electrical wiring.

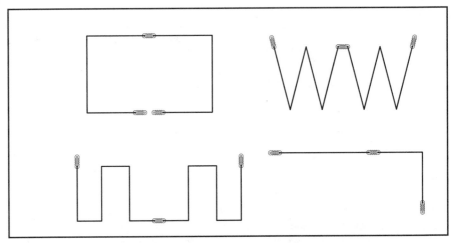

Figure 7-9—You can bend dipoles into all kinds of creative shapes to fit your yard, room, or attic.

Loop antennas like this one are ideal for indoor use. The MFJ model 1786 offers 10 to 30-MHz performance in an antenna that's only 3 feet in diameter.

Never say never! Bob Derbacher, W6REF, installed this 10-meter antenna inside his townhouse. When Bob's not using it as an antenna, it becomes a curtain rod!

Ladder-line fed dipoles are ideal for attic use—assuming that you can route the ladder line to your radio without too much metal contact. In the case of the ladder-line dipole, just make it as long as possible and stuff it into your attic any way you can. Let your antenna tuner worry about getting the best SWR out of this system.

For small attics you may want to consider a small tunable *loop* antenna, such as one of those manufactured by MFJ Enterprises. These incredibly small antennas offer decent performance on all but the lowest HF bands. They're basically thick circles of metal that are tuned by large, motor-driven capacitors. By adjusting a control at your radio, you change the capacitor setting and, as a result, the resonant frequency of the loop.

The same dipoles and loops that you use in your attic can also be used in any other room in your home. The same techniques apply. Keep the antenna as high off the floor as possible. (As with most antennas, the more height, the better.) For indoor operating, however, I recommend using low output power. You'll avoid RF "bites" as well as interference to VCRs, TVs and so on. Many hams have been successful operating indoor antennas with just a few watts output.

Accessories

Now that you've purchased your transceiver and antenna system, it's time to take a quick look at some station accessories. If you still have some disposable income nestled in your bank account, consider the following adornments for your operating area . . .

• **SWR meter**

SWR meters provide visual indications of the impedance match between your transceiver and your antenna system. They measure the magnitude of forward and reflected power, displaying the result as a *standing wave ratio* (SWR). Generally speaking, an SWR of 2:1 or less is acceptable to modern solid-state transceivers. At SWRs greater than 2:1, most transceivers activate a "foldback" circuit to prevent damage to the output transistors. The foldback curtails the RF output power, dropping it precipitously depending on the severity of the SWR.

Many rigs include an SWR meter. Make sure yours doesn't before you make an extra purchase! And if you intend to buy an antenna tuner (see below), you'll be pleased to know that most tuners include their own SWR meters.

Obsessive creatures that we are, most hams

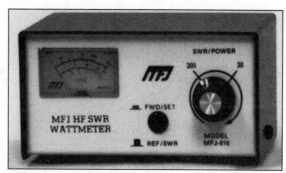

This innocent-looking box is one of the most essential pieces of test equipment in your HF station: the SWR meter.

prefer to monitor their SWR at all times. A screwy SWR reading tells you right away that something is wrong with your antenna system. (If an overnight ice storm coats your antenna, you'll see the result as an elevated SWR the following morning.) Modern SWR meters also display your average output power, which is another nice thing to know.

So, if your radio doesn't include an SWR meter, and you don't need an antenna tuner, sink about $100 into a good stand-alone meter anyway. You'll be glad you did.

• Antenna Tuner

Consider this an adjustable impedance matcher between your antenna system and your transceiver. An antenna tuner doesn't really "tune" the antenna in a literal sense.

Antenna tuners come in all sizes according to how much RF power they can handle and the designs of their matching circuits. Prices range from about $80 to well over $1000. If you're using the typical 100-W transceiver, you should be able to find an adequate tuner in the $120 to $200 price range.

If you intend to use an antenna fed with ladder line, or a random-wire antenna, you *must* have an antenna tuner. For coaxial-fed antennas, however, tuners are optional. If the SWR at your operating frequency is less than 2:1 on a coax-fed antenna, you don't need a tuner. (Besides, some HF transceivers feature built-in tuners.)

• Keys and Keyers

If you want to operate CW, you'll need a CW key. So-called "straight" keys can be yours for a minimal price in some cases, but they're tiresome to operate during long conversations (or contests). "Paddle" keys are used with electronic keyers. The keyers automatically generate the dits and dahs according to which paddle you're pressing at the time. With a paddle key and keyer at your side, you can send copious

Antenna tuners come in all shapes and sizes. An economy tuner such as the MFJ-949E will match random wires, ladder-line fed dipoles, or coax-fed antennas.

The Nye Viking MB-V-A is a "high-end" antenna tuner designed to match just about anything. Although it's expensive, the tuner is built like a tank and can easily handle 1500 W.

In this MFJ unit, the paddle key *and* the keyer coexist. The controls on the side adjust dot-dash speed and sidetone volume. You can create "canned" CW messages and store them in memory. Just press one of the top-cover buttons to play them back.

This headset gives you total hands-free operating. The microphone is part of the headphone assembly. Swing the microphone down in front of your lips and transmit!

amounts of code with slight movements of your thumb and index finger. Electronic keyers are considered required equipment these days by serious contesters and DXers.

• **Headphones and Headphone/Microphone Combos**

Headphones are a blessing to any ham who lives with someone, or has crabby neighbors. Radio noises are music to our ears, but they often drive others up the nearest walls. A good set of headphones is a terrific investment for domestic tranquillity. They're also wonderful aids when working those weak stations you can barely hear. You'll be amazed at how much more you can copy with the audio output of your radio in both ears.

If you want to enjoy the luxury of hands-free voice operating, consider a headphone/microphone combination. The microphone is mounted on a small boom attached to one of the earpieces. The boom is adjustable, allowing you to position it right in front of your lips. DXers and contesters love these gadgets! They can rack up the contacts and keep their hands free to type, write, eat, wave their fists in the air, or whatever.

• **Logs**

Although FCC logging requirements are a quaint memory, hams still enjoy keeping track of their contacts. A well-maintained log is like a diary, it allows you to step back in time and rehash some memorable conversations. Logs have more practical uses, too. If someone accuses you of interfering with their television, for example, you can consult your log and see if you were actually on the air at the time. Logs can also help you get a handle on how your antenna system is performing. Do most of your contacts seem to come from particular areas of the country? If so, perhaps your antenna radiates your signal most strongly into those areas.

Personal computers are ideal machines for logging. Check the pages of *QST* and you'll

Antenna Tuner—Do You Need One?

Buy an Antenna Tuner if . . .

. . . you want to feed your antenna with open-wire line.

Open wire line (or *ladder line*) offers extremely low loss at HF frequencies (much better than coaxial cable). One problem is that open wire line is *balanced* while your transceiver output is *unbalanced*. You need to use an antenna tuner with a built-in *balun* to form a bridge between the balanced line and the unbalanced output of your radio. A balun is a type of transformer that converts balanced feed lines to unbalanced, or vice versa. (*BAL*anced to *UN*balanced. Get it?) Most antenna tuners use 4:1 baluns, which also convert the impedance of open-wire feed lines to a value that the tuner can handle.

. . .you want to operate your antenna on bands other than those it was designed for.

When you attempt to use, say, a 40-meter dipole on 10 meters, a big mismatch will develop along with a high SWR. By using an antenna tuner, you may be able to create a 1:1 SWR at your transceiver. (I say "may" because the mismatch can often be so great that it is beyond the capability of your tuner to handle.) The high SWR may cause substantial loss in a coaxial feed line, but at least you'll radiate *some* power at the antenna.

. . . your antenna has a narrow SWR bandwidth on some bands.

Some types of multiband antennas do not offer low SWRs from one end of each band to the other. There is usually a range—expressed in kilohertz—where an SWR below 2:1 can be achieved. For example, a multiband trap dipole may offer an SWR of 2:1 or less from 3600 to 3800 kHz. That's an SWR bandwidth of 200 kHz. If you try to operate above 3800 kHz or below 3600 kHz, you'll encounter an SWR higher than 2:1 and your radio may become displeased. With an antenna tuner, you can operate outside the SWR bandwidth and still load the full output of your radio into the antenna system.

Don't Bother With an Antenna Tuner if . . .

. . . your SWR is 2:1 or less at the frequencies you operate most often.

An SWR of 2:1 or less is not serious and does not require the assistance of an antenna tuner. Most modern rigs will tolerate a 2:1 SWR just fine. If you are using a good-quality feed line, the loss caused by an SWR of 2:1 isn't enough to worry about at HF frequencies. Many hams are obsessed with providing an absolute 1:1 SWR for their radios at all times. Apparently they also have money to burn!

. . . you're interfering with TVs, telephones and other appliances in your neighborhood.

Despite what you may have heard, an antenna tuner will not necessarily cure your

find software that maintains your log, prints QSL address labels, and does just about everything else except brew your coffee.

• **QSL Cards**

A QSL card is written verification of a contact. It contains all the vital information such as the time of the contact, the frequency, signal reports and so forth. Hams who are chasing various awards primarily use QSLs. They need QSLs to prove that the contacts were genuine. Even if you're not seeking certificates to hang on your wall, it's a good idea to keep a stock of cards at hand. After all, *you* may be the contact someone needs to clinch their award! You'll find plenty of QSL printers advertising in the back pages of *QST* magazine. Most will send you samples of their work for a small fee.

interference problems. It's true that an antenna tuner may reduce the level of *harmonic radiation* (signals your radio generates in addition to the ones you want). If the interference is being caused by harmonics, a tuner may help. However, most interference is caused by RF energy that's picked up indirectly by cables or wires, or directly by the device itself. By using an antenna tuner, you'll probably radiate more energy at the antenna than you did before. That may make your interference problem worse!

Looking for Mr Goodtuner

So, you've decided that you need an antenna tuner after all. Antenna tuners come in all shapes and sizes. What features should you consider?

• A built-in SWR meter

An SWR meter of some type is a must if you want to use an antenna tuner. When adjusting your tuner, you need to keep your eye on the *reflected power* indicator. Your goal is to reduce the reflected power to zero—or at least as close as you can get. When the reflected power is zero, the SWR is 1:1 at your transceiver.

• A roller or tapped inductor

More expensive tuners feature a variable coil called a *roller inductor*. As you turn the front-panel inductor knob, the coil inside the tuner rotates. A metal wheel rolls along the coil like a train on a railroad track. As the wheel moves along the coil, the inductance increases or decreases.

Some tuners do not use roller inductors. Instead, there is a coil with wires attached at various points. On the front panel, a rotary switch selects the wires. According to how the inductor is wired in the circuit, selecting one *tap* or another varies the inductance. This is known as a *tapped inductor*.

There are advantages and disadvantages to both approaches. Roller inductors offer the best tuning performance, but they are subject to the woes of mechanical wear and tear. For example, if corrosion builds up on the wheel or the coil windings, the electrical quality of the connection will deteriorate. Roller inductors are also cumbersome to use. You may have to twist the control many times when moving from one band to another.

Tapped inductors are easy to use and free of mechanical problems (unless the switches get dirty). However, you may find that they restrict the operating range of your tuner. When you turn the switch, you select a *fixed* amount of inductance. You can't easily change it to tune a particularly difficult mismatch situation.

• A built-in balun

If you ever intend to use an open-wire feed line, buy a tuner with a 4:1 balun built-in.

(continued on next page)

A catchy QSL card is sure to get noticed in a pile of mediocre confirmations!

These baluns often dissipate quite a bit of heat, so always choose a large balun over a small one.

• Multiple antenna capability and dummy loads

Some tuners offer the ability to connect more than one antenna. This is handy in all sorts of applications. Let's say you have a vertical antenna for 40-10 meters and a wire dipole for 80 meters. You can connect both feed lines to your tuner and easily switch between them.

Built-in dummy loads are convenient, but not necessary. A dummy load is a resistor (or group of resistors) that absorbs the output of your transceiver while allowing very little energy to radiate. It's used for making transmitter adjustments and other tests. If your tuner lacks a dummy load, you can purchase one separately.

A Word about Power Ratings

If your transceiver produces only 50 or 100 W of power, a 200 or 300-W tuner should do the trick, right? Well . . . yes and no. A high SWR can result in high RF voltages in the tuner. If you're trying to use your tuner in a high-SWR situation, the RF voltages at the tuner may cause an unpleasant phenomenon known as *arcing*. That's when the RF energy literally jumps the gaps between the capacitor plates or coil windings. When your tuner arcs, you'll usually hear a snapping or buzzing noise. The reflected power meter will fluctuate wildly. Interference to your TV and other devices will increase dramatically. You may even see brilliant flashes of light inside your tuner!

Arcing is obviously bad news for your tuner. It's your tuner's way of saying, "Stop! I can't handle this mismatch!" There are only two cures for arcing: reduce your output until it stops, or get a tuner with a higher power rating.

High-power tuners use large capacitors and coils. The gaps between the plates and windings are greater, making it more difficult for an arc to occur. If you can afford it, you're always better off buying a tuner with a 1.5 kW rating or better. A hefty tuner costs more, but it will serve you well in the long run.

Buy or Build?

As you comb through the advertising pages of *QST*, you'll see many new antenna tuners for sale. The prices are often reasonable and the quality is usually good. Keep your eyes open for used tuners, too. If a used tuner is in decent condition, it's every bit as usable as a new one.

If you like to build things, however, consider an antenna tuner as your next project. Antenna tuners are very easy to construct. You can find capacitors and coils at hamfest flea markets at very low prices. Even roller inductors—the most expensive part of a roller-inductor tuner—can be found for less than $40 if you look carefully.

Your chances of success with an antenna tuner project are excellent. You have to try pretty hard to build one poorly! Best of all, you'll have the satisfaction of using a piece of equipment that you've put together yourself. *The ARRL Handbook* offers several tuner designs you can try. Heat up your soldering iron and go to it!

SCARING UP CONTACTS

When you've assembled and tested your station there is nothing left to do but get on the air. I recommend several hours of *listening* before you reach for the key or microphone. Just tune through the bands and eavesdrop on as many conversations as possible. When you finally feel comfortable with the territory, it's time to throw some RF!

CW

The best way to start a CW chat is to tune around until you hear someone calling CQ. CQ means, "I wish to contact any amateur station." In time you'll learn to recognize the sound of a CQ call. It has an unmistakable rhythm!

If you can't find anyone calling CQ, perhaps you should try it yourself. A typical CQ goes like this: CQ CQ CQ DE KD4AEK KD4AEK KD4AEK K. The letter K is an invitation for any station to reply. If there is no answer, pause for 10 or 20 seconds and repeat the call.

If you hear a CQ, wait until the ham finishes transmitting (by ending with the letter K), then call him. Make your call short, like this: K5RC K5RC DE K3YL K3YL AR (AR means "end of message").

Suppose K5RC heard someone calling him, but didn't quite catch the call because of interference (QRM) or static (QRN). Then he might come back with QRZ? DE K5RC K (Who is calling me?).

The Conversation is Underway

Most HF contacts begin with an exchange of basic information: Names, locations, equipment, signal reports and even weather reports. After that, it's up to you. Sometimes you'll find that you have to draw the other person into the conversation. The best way to do that is to ask questions. For example, ask what the person does for a living. She's a doctor? Okay, ask about her specialty, where she

---

What If No One Is Calling CQ?

You're just spent 15 minutes tuning up and down the band. You're dying to talk to someone, but you can't find anyone calling CQ. Should you pull the big switch and try again later? Not necessarily.

Stop and listen to some of the conversations in progress. With luck you'll find one that is about to end. Now listen *very* carefully. Assuming that you can hear both sides of the chat, one person is probably saying that they have to run to the store, the post office, the bathroom or another urgent errand. Scratch that guy. He doesn't want to linger. But what about the other operator? Is he or she just signing off, in no particular hurry to be anywhere? There's your target! As soon as the conversation ends, give that person a call. About 50% of the time you'll find the other operator is more than happy to start a new round of discourse.

RST

Every ham wants to know how well his or her signal is reaching your location. It's a point of pride! So, the best thing you can do is to give them an honest evaluation—by using the *RST* system:

R — Readability

S — Strength

T — Tone

For readability choose a number from 1 to 5. For tone (used on CW and the digital modes only) and strength, select a number from 1 to 9.

Readability:

5 = Perfectly readable

4 = Readable with practically no difficulty

3 = Readable with considerable difficulty

2 = Barely readable, occasional words distinguishable

1 = Unreadable

Strength:

9 = Extremely strong

8 = Strong

7 = Moderately strong

6 = Good

5 = Fairly good

4 = Fair

3 = Weak

2 = Very weak

1 = Faint

Tone:

9 = Perfect

8 = Near perfect

7 = Moderately pure, just a trace of distortion

6 = Some distortion

5 = Moderate distortion

4 = Rough note

3 = Very rough note

2 = Harsh and broad

1 = Extremely harsh

It's up to you to select the numbers that best describe the signal you're hearing. An SSB signal that blows your doors off is obviously 59. But another SSB signal can be weak, yet perfectly readable. That signal might earn a 55 or even a 53.

The same ideas apply on CW and the digital modes. However, you have the tone factor to consider. In this era of modern transceivers, it's rare to hear a CW or digital signal that rates anything less than a 9 in the tone department. But if you hear a signal that bears the raspy signature of, say, a failing power supply, let the fellow know by giving him an appropriate rating.

practices and more. In other words, get her to talk about herself. If you ask the right questions, the conversations will unfold on their own.

During the contact, when you want the other station to take a turn, the recommended signal is KN, meaning that you want only the contacted station to come back to you. If you don't mind someone else signing in, just K ("go") is sufficient. You *don't* need to identify yourself and the other station at the beginning and end of every transmission. That wastes time. The FCC only requires you to identify *yourself* every 10 minutes.

Ending the Conversation

When you decide to end the contact, or when the other ham expresses his/her desire to end it, don't keep talking. Briefly express your thanks: TNX QSO or TNX CHAT—and then sign out: 73 SK WA1WTB DE K5KG. If you are leaving the air, add CL to the end, right after your call sign.

SSB

To get an SSB chat off the ground, you have two choices: You can call CQ, or you can answer someone who is calling CQ.

Before calling CQ, it's important to find a frequency that appears unoccupied by any other station. This may not be easy, particularly in crowded band conditions. Listen carefully—perhaps a weak DX station is on frequency.

No matter what mode you're operating, *always listen before transmitting*. Make sure the frequency isn't being used *before* you come barging in. If, after a reasonable time, the frequency seems clear, ask if the frequency is in use, followed by your call. "Is the frequency in use? This is NY2EC." If nobody replies, you're clear to call.

Keep your CQ very short. Longwinded CQs drive most hams crazy. Besides, if no one answers, you can always call again. If you call CQ three or four times and don't get a response, try another frequency.

A typical SSB CQ goes like this:

"CQ CQ Calling CQ. This is AD5YER, Alfa-Delta-Five-Yankee-Echo-Romeo, Alfa-Delta-Five-Yankee-Echo-Romeo, calling CQ and standing by."

And if you're the caller (as opposed to being the "callee"), keep the call short. Say the call sign of the station called once or twice only, followed by your call repeated twice.

"N2EEC N2EEC, this is AB2GD, Alfa-Bravo-Two-Golf-Delta, Over."

Chewing the Rag

"Rag chewing" is ham lingo for a long, enjoyable conversation. As with CW contacts, start with the basic facts: your name, location, his signal report, and possible a brief summary of your station (how much power you're running and the kind of antenna you're using).

Once you're beyond the preamble, the topic choice is yours. The tried and

Operating QRP (low power, defined by the ARRL as 5-W output or less) is a popular modus operandi of thousands of hams. The thrill of communicating at low power levels is perhaps best described as being similar to the excitement experienced during your first QSO. Interestingly enough, the level of enjoyment proportionately goes up as your power goes down.

The first thing anyone contemplating a jump into the QRP sport should do is cast off the notion that you must run high power (QRO). A more enlightened attitude is needed. Consider your QRP operating as an adventure, a challenge, a unique and very personal voyage on the airwaves—riding a leaf instead of a supersonic jet. It's a gentle form of communication; think "heart and soul," not "blood and guts." Shoot down your DX prey with a peashooter rather than a double-barrel shotgun. A positive frame of mind will set the stage for an enjoyable time with QRP. Here are some important ground rules:

1. Listen, listen, listen.
2. Call other stations, don't call CQ.
3. Expect less-than-optimum signal reports.
4. Be persistent and patient.
5. Know when to quit.

Listen to the bands and try to figure out what the prevailing propagation is. Is the skip short or long? Who's working who? Is there much interference, static or fading (QRM, QRN or QSB, respectively) present? A quick analysis of the band conditions should be the first thing you do when sitting down for a session of QRP operating. Listening will help you decide what band to operate. Always listen, listen, listen!

Call CQ as a last resort! Most hams prefer to answer a strong signal, which you probably will not have. You will be much better off answering a CQ. Try answering someone like this: WJ1Z DE W1RFI/QRP or WJ1Z DE W1RFI/2W K. This tells WJ1Z why you aren't doing a meltdown on his headphones!

Don't be discouraged if you receive signal reports like RST 249 or RS 33. With less than 5-W output, you can't expect to be overloading the receiver front ends out there in DX-land. Have faith, for you will get more than your fair share of very respectable reports. The ultimate ego gratification of a 599 or 59 will be yours if you keep at it!

If at first you don't succeed, then try again. And again. This QRP stuff is a game of persistence, so don't give up if you don't get an answer to your call on the first try. Don't be discouraged. Make up your mind—instant success just isn't part of the plan—that's what makes it so much fun.

---

true formula for success is to get the other person to talk about himself. Any life worth living has at least *one* interesting aspect. You may have to dig this aspect out of your palaver partner, but it's often worth the effort. If all else fails, make the following request:

"Look out the window and tell me, in detail, exactly what you see."

You'll definitely throw the other person off guard—that much is guaranteed! If they're in a room without a window, don't let them off the hook. "What would you see if you *did* have a window?"

A very useful tool for the QRP station is the wattmeter. A commercially built unit can be found for a reasonable price. A basic single meter unit with switchable forward and reverse power is a good way to start. In time you may want to add another meter to eliminate the need to switch back and forth between forward and reverse power. Save the switch and use it to change the power range of the meter. This way you can have one range for a 5-W full scale, and the other a 1-W scale. You can calibrate this wattmeter with a VTVM (vacuum-tube voltmeter), a simple homebrew dummy load and an RF probe. Not only will you be saving some hard-earned bucks, but you will be gaining experience in designing, building, modifying and calibrating test equipment.

With QRP, your antenna is going to be much more important and instrumental in your success than if you run QRO. Running 100 W into a random wire will net you plenty of solid contacts. But when you reduce power into the same wire, your signal effectiveness will decrease, too. As a result, the old axiom of putting up the biggest antenna you can muster, as high and as in the clear as possible, means more to the QRPer than someone running 100 or 1000 W. The important thing is to optimize your antenna to your own personal circumstances. Many operators have reported amazing results using less-than-optimum antenna systems, but this is not to say you should be lax in your antenna installation. By running QRP, you are already reducing your effective radiated power (ERP); no need for a further (unintentional) reduction by cutting corners on your antenna system.

For 160 to 30 meters, dipoles generally will work well. Height is always nice, so do the best you can. Loops, end-fed wires and verticals are also used on these bands. The popular HF DX bands, 20 to 10 meters, deserve some serious thought as to rotatable gain antennas—Yagis or quads. Although this train of thought usually leads to a considerable outlay of cash, you will benefit in several ways. A 1-W signal to a 10-dB gain Yagi will give you an effective radiated output of about 10 W! That's just like having an amplifier that needs no power to run. A directional antenna is a reciprocal device as well. It is effective on both received signals and the transmitted signals. Listening to Europeans is so much more fun when you don't have to hear them along with signals from other unwanted directions.

On the bands, CW QRP activity centers around 1810, 3560, 7040 (7030 for QRP DX), 10,106, 14,060, 21,060 and 28,060 kHz. Voice operation is around 3,985, 7,285, 14,285, 21,385 and 28,885 kHz. The 10, 18 and 24-MHz bands are also hotbeds of QRP activity. Novices should check 3,710, 7,110, 21,110 and 28,110 kHz. —*Jeff Bauer, WA1MBK*

DXING 'TIL DAWN

Strictly speaking, DX is any contact that you make with a ham in another country. But real DX is in the eye of the beholder. If you're new to the HF bands, DX is the long chat you just had with a fellow in Great Britain. To an HF veteran, however, DX is a five-second contact with the only amateur in the Republic of Chad.

Many hams compare DXing to fishing. You can spend minutes, even hours, drifting across the bands, listening to one conversation after another. Then,

(continued on page 7-32)

The Importance of *Zero-Beating*

By Bob Shrader, W6BNB

Very simply stated, zero beating is the process of one ham tuning his transmitter to exactly the same frequency as that of a station he is receiving. The beat note (ie, the frequency difference) of the two stations is zero. When you listen to two hams talking on SSB, they are zero-beat if both their voices sound normal, as you listen without your touching any controls on your receiver (or transceiver). When you listen to two hams talking on CW, they are zero-beat if you hear the note of both transmitted signals at exactly the same audio pitch, again without your touching any of the controls on your receiver (or transceiver).

It is simply good operating practice to have both hams in a QSO on the same frequency, so as not to spread out and take up a wider portion of the band than is necessary. This becomes even more important in net or roundtable operation, where several CW stations who are mistuned by a few hundred hertz can cause the bandwidth in use to be perhaps two to five times that which would be used if the stations were zero-beat. Furthermore, it causes you to be continually adjusting your RIT (receiver incremental tuning) control to copy each of the different stations.

Are you beginning to see the importance of zero beating? Well, then, let's see what's happening on the air these days.

Our HF bands are often full of signals, so we need to work carefully to squeeze the greatest number of stations possible into the limited spectrum. How? Well, we can reduce power to the minimum necessary to make the contact—which may help a bit. As a matter of fact, there is an FCC rule that requires that this be done! We can use rotary beam antennas, so as to minimize interference to stations in directions other than the "beamed" direction, which also minimizes the interference received from all the other directions. And (drum roll and cymbal crash) we should always accurately zero-beat the station we're talking with, so as to minimize the bandwidth used for the contact!

"How To Do It" for SSB Enthusiasts

First make sure your receiver incremental tuning (RIT) control (sometimes called OFFSET) is turned off (or, if it doesn't have an off position, the RIT must be set to the middle of its range). With the RIT control turned off, your transmitter and receiver automatically transmit and receive on the same (carrier) frequency. To zero-beat a signal, all you have to do is to listen carefully to the voice characteristics of the signal you're tuning in. Adjust your frequency control to make the voice sound as natural as possible. Listen to all the components of the voice; you don't want to hear any strange low-pitched components or any garbled high-pitched components. Practice tuning in SSB stations with this thought in mind, and you will soon develop the preference for properly tuned signals, and a corresponding intolerance for the strange sounds of improperly tuned signals. This is to ensure that your carrier frequency is zero-beat with the other station's carrier frequency all of the time.

And then there is the older SSB equipment, where you have a separate transmitter and receiver. First, tune your receiver for the naturalness of voice that was just described. Then turn your transmitter's microphone gain down to zero and key up the transmitted signal. Tune your transmitter's frequency control until you start to hear a whistling sound, and continue tuning until the whistle goes to its lowest-frequency tone and, finally to no whistling sound at all. Voila! Your transmitter is now zero-beat with the received station. You won't interfere with other stations that might be listening to the station you are zero-beating; with SSB transmitters, the carrier frequency is suppressed and nothing is transmitted until you start talking. By the way, remember to reset your mike gain to its proper operating position after you have finished zero-beating, or you won't be transmitting when you turn on the rig and start talking!

But CW Is Another Story!

Let's start by figuring out what to do if you have a separate transmitter and receiver. Tune your receiver to the station you want to zero-beat, with the audio pitch (or tone) of the signal set to the pitch you like to copy (this varies from individual to individual, so suit yourself). Reduce your transmitter output to its lowest level (or use the spotting function if your transmitter has one). Now adjust your transmitter frequency so that you hear your own signal, when you tap out a few dots, at the same audio pitch as that of the other station. There you are: zero-beat! After you've done it a few times, you'll get the hang of it and can do it very quickly.

But if you're using vintage equipment, you may not have single-signal reception. With single-signal reception, as you tune through a CW signal, you will hear little or no signal on one side of its center frequency. Without it, you will hear the signal come into the audio band pass as a high-pitched audio note as you're tuning through, decreasing to zero, then going back up in pitch on the other side. Therefore, without single-signal reception, when you try to zero-beat you can inadvertently tune your transmitter to the "wrong" side of the other station's signal. When this happens you will be pretty far off frequency; if you are copying the other signal at a 700-Hz audio pitch, then you will be twice that amount—or 1400 Hz—away from being zero-beat. If you're that far off frequency, the other station will most likely not hear your call!

Now let's consider zero beating with the common transceiver. At first, this might seem dead easy. Not so! Different manufacturers use different frequency offsets between the received and the transmitted signal, and sometimes a given manufacturer will have different offsets for different models in his product line.

Virtually all transceivers, when the RIT control is turned off, will transmit on a frequency that is between 500 and 1000 Hz different than the frequency being received; this is called the frequency offset. The concept is that if you tune your transceiver to copy a received signal at an audio pitch that is equal to the offset frequency (typically about 800 Hz), you will automatically be zero-beat. Sounds good—right? But what if you prefer to copy at an audio pitch of 500 Hz? Then when you tune your transceiver frequency to the audio pitch you prefer to copy, you will be 300 Hz (ie, 800 – 500) away from zero-beat—every time! What to do?

One of the best ways to learn to zero-beat other stations is to enlist the help of two other hams—experienced CW operators, if possible. Make some arrangement so you can all communicate either on another band, such as 2-meter simplex, or via a conference call on the telephone. In the following discussion, you will play the role of Station 1, and you will attempt to zero-beat Station 2, while Station 3 listens to critique your efforts and guide you to being zero-beat with Station 2.

The three of you will talk on your alternative frequency (2 meters or whatever), and Station 2 will transmit on some frequency, say about 7.050 MHz. Once Station 3 (the observer) has tuned in Station 2 on his receiver, he will tell you to try to zero-beat Station 2. You tune to what you think is zero-beat, and then Station 3 reports to you which direction (up in frequency or down) and about how far (1 kHz, 500 Hz, 200 Hz, or whatever) you need to move to be zero-beat. After a few such directions, Station 3 will eventually get you zero-beat with Station 2.

Now the important work begins: At this point, you need to listen carefully to the audio pitch of Station 2's signal—that's the audio note at which you must receive another station,

(continued next page)

The Importance of Zero-Beating (continued from page 29)

in order to be zero-beat. To "remember" a specific tone is almost impossible, except for the few people with the musical talent known as "perfect pitch." You could use an external audio oscillator with an adjustable output pitch, setting its pitch to match the audio pitch you are receiving from Station 2. Then, to zero-beat another station, all you need to do is to tune the transceiver's frequency until the received signal's pitch matches your preset reference pitch from the audio oscillator.

Another possibility is to use either an audio frequency tuning control, if you have one, or an external narrowband audio filter set so that its peak frequency matches the pitch of the signal from Station 2. Using this technique, you would then tune in a received signal until the volume of the received signal peaks. (The inherent accuracy on this method is not as good as that of matching the audio pitches, but it's usually good enough to get you within 100 Hz of zero-beat.)

Well, You Get the Idea

All of this may sound complex, but don't be discouraged. A little practice will enable you to learn to zero-beat other stations accurately—a good target is always to be within 30 Hz of true zero-beat. This truly is an important concept, so work at developing your zero-beating skill; when you've mastered it, you will be a more competent operator, and the skill will increase your ability to make and hold contacts on the air.

Which is the Best Mode for DX? SSB or CW?

During periods of above-average propagation conditions, Novices and Technicians will find more DX on 10-meter voice than on either of their 15 or 10-meter CW subbands. But, as go the sunspots, so goes 10 meters. When solar activity is low, the 15-meter Novice/Technician CW subband is more productive. Novices and Technicians should not overlook their CW subbands when the sun is active either, because some choice DX occasionally shows up there. (Beginners in other countries have to study the code, too.) And, because activity in the Novice 15 and 10-meter CW subbands is lower than in the 10-meter voice band, there is less competition and a greater chance for a QSO. Further, if you are really interested in DXing, you need a higher class of license and the ability to sling out the dits and dahs.

This opinion is not just the raving of an old-time CW nut! I could fill three books with the gory details of DXpeditions and other rare ones I have called for hours on SSB, when I just couldn't break the pileup. When the DX finally got on CW, I often got through on the first call. SSB receivers use wider bandwidths by necessity and all the other strong signals calling simply overload the automatic gain control (AGC) circuit. CW allows the use of much narrower filters. Also, a good CW operator develops the ability to separate signals in his brain. This ability is no mystery, because with CW we're dealing with single tones, not a mish-mash of voices.

If you have a General class license or higher, you can operate on the other HF bands not available to Novices and Technicians. There, you should also concentrate on CW, especially if you don't have a big signal. True, some of the rarer DX stations sometimes operate in the Amateur Extra Class subbands, but a majority of those operators listen higher in the band, where General and Advanced class amateurs can transmit. If you get hooked on DXing, the first time you hear a really rare station working in the Amateur Extra Class subband, you'll be surprised how fast you can get your code speed up to snuff! Meanwhile, you can readily work more than 100 countries without going into the Amateur Extra Class subband.

What about SSB? True enough, there is plenty of DX on SSB. Almost every country I hear people fighting over on SSB, however, I have easily worked on CW (and usually on more than one band). Enough said. CW is a different means of communication and no one masters it overnight. Most of us old-timers couldn't wait to get on voice (in the days when Novices were limited to CW). Many of us found our way back to CW again, though, simply because we like it better. No matter how wild the pileup may be on CW, it just doesn't sound as bad as even a low-key pileup on SSB. There is always something happening in the Amateur Extra Class voice subbands and you won't want to miss it.

Why is there so much DX on CW? Most American Radio amateurs can afford a rig that works both modes, but this isn't true in many foreign countries. In many parts of the world, amateurs can't buy commercial equipment at any price and obtaining crystal filters and other parts necessary to build SSB gear is difficult, if not impossible.

You won't work these folks on any other mode. CW is harder at first; you know how to speak with your voice—now you must learn to speak with your key. I am not suggesting you give up SSB. Quite the opposite: I am hoping you won't give up CW. Being flexible is the mark of a true-blue DXer.

Getting By on CW

Let's say you decide to take the plunge and you're listening on a CW subband. Suddenly, you hear a bunch of stations sending their calls for a few seconds, spread out over several kilohertz. Then, just as suddenly, all is quiet. You tune lower in frequency and hear a weaker signal sending "5NN" ("N" is an abbreviation for "9"), but it sounds more like a machine-gun burst. There's your DX station. Who is it? Stick around, you'll hear the call sign soon enough.

Now he's calling another station, but you can only copy part of the call sign because he's going so fast. If it were your call sign he was sending, would you recognize it? If not, no sense adding to the QRM by calling him. Find another station. But for now, let's say you can follow what he's sending. There he goes again, he just sent "TU." That's his way of acknowledging the other station's report. He's listening up the band for another call now—yes, he just let loose with another burst. Wow, you really have to concentrate!

Why not listen above him, and see if you can find the station he's working? There's someone else, sending at the same machine-gun speed, sending his call at the end of his report. Okay, back to the DX station again. A few contacts later, you've got his routine figured out and you can copy most of what he's sending.

Assuming your license allows you to transmit on this frequency, why not give him a shout? You know about how long he listens before he picks out a call. And you know whether he tunes up, down, or randomly after each contact Here's a hint: If you send your call a little slower than he's sending, he might slow down when he calls you. Sometimes a slowly sent call sign stands out in a pileup of fast callers, too.

Okay, you say, here goes "KR1S." Nope, someone else got him. Let's see, the station he called was a little lower in frequency than me. Guess I'll stand pat for one more call. Here we go again. "KR1S." Yikes! He just called me! Gosh, "5NN de KR1S TU." He sent "TU." I guess he heard me. Not exactly long winded, is he! Oh, yeah, what was his call sign? I figured it out a few minutes ago and wrote it down . . . here it is. Okay, put it in the log and don't forget the UTC time and date. Hey! I just worked a new one on CW! Gee, that wasn't too bad. Where'd I put that W1AW code-practice schedule, anyway?— *From* The DXCC Companion, *by Jim Kearman, KR1S* suddenly, there's a "tug" on your line. What's that commotion? Who are all those stations calling? You listen for a lull in the RF storm. Yes, there's the DX station. Holy cow! It's Heard Island!

Now the fish has taken the bait. Your bobber just dipped below the murky surface, but the prize is far from hooked. You have to dive into the fray and do battle for the attention of the DX station.

It could take an hour or more to land your prey, or you may not land him at all. That's the thrill of the hunt. When you hear the DX station finish with one contact, it's your turn to call. In a pileup (where many stations are calling at once) it's often best to simply send (or say) your call sign once. Now listen. Has he answered anyone? If not, send your call and listen again. As soon as you hear the DX station begin a new contact, stop calling. Yes, he might hear you in the background while he's trying to copy the other station, but you'll achieve nothing other than making him angry. He might even jot down your call sign—along with a reminder to *never* answer your call!

Prepare yourself for a jolt of giddy excitement when you hear the DX

DX QSLing: Bureau or Direct?

QSL bureaus are useful if you're a DXer on a budget. Think of them as small post offices. As an ARRL member you have exclusive access to our Outgoing QSL Service. For just 10 cards for $1 or $4 per half pound or portion thereof [at this writing], they'll ship your QSL cards to bureaus overseas. (A pound is a *lot* of QSL cards!) The overseas bureaus, in turn, will send your cards to the DX stations.

To use the outgoing service, presort your cards according to the call-sign prefixes of the destination DX stations. Enclose an address label from your current copy of *QST* (to prove membership), along with a check or money order. Wrap the package securely and send it to:

ARRL Outgoing QSL Service
225 Main St
Newington, CT 06111

Working in reverse, the DX stations send cards to their local bureaus, who in turn ship them to *incoming* QSL bureaus here in the US. (Each call-sign district is served by an incoming QSL bureau.) If you have a self-addressed, stamped envelope (SASE) on file with the bureau that services your call district, they'll send the cards to you right away.

By the way, send your SASE to the incoming bureau that serves the district according to your call sign—even if you've moved to another district. Let's say you lived in Alabama and were issued the call sign AA4XYZ. A few years later, you moved to Idaho (the seventh call district), but you kept your four-land call. You would still keep your envelope on file with the district-four bureau! Of course, if you switched to, say, KC7XYZ, you'd also switch to the district-seven bureau.

The bureau system is popular because it is so inexpensive. The drawback is that the system is very slow. It's common to wait a year or longer to receive a card via the bureau. That's why some hams prefer to mail their QSLs directly to DX stations. They usually include $2 for return airmail postage and a self-addressed airmail envelope.

operator sending *your* call sign. Yes, he heard you! You're in the spotlight—for the next few seconds anyway.

On SSB it may go like this:

"NØSEJ you're 59." (Or he may just use the suffix of your call sign: "SEJ you're 59.")

As you choke back your joy, you reply, "JT1DHF from NØSEJ. You're also 59. Thanks and 73!"

On CW the same exchange might be:

NØSEJ 599 TU ("TU" means "to you," as in "back to you")

JT1DHF DE NØSEJ QSL 599 TNX ES 73 (Translation: JT1DHF from NØSEJ. Understood. You're also 599. Thanks and 73.)

That's it! You're done and he's already working someone else. Congratulations!

A "Split-Decision"

Many DX stations choose to "work split." That is, they transmit on one frequency and listen for calls on another frequency (usually a group of frequencies). For example, the DX operator may transmit on 14.195 MHz and listen for calls from 14.210 to 14.220 MHz. To snag a split-frequency DX contact you'll need a radio with two VFOs. Fortunately, most modern transceivers include this feature. The trick is to make absolutely certain that you're not transmitting on the frequency where the DX station is transmitting. You want to *listen* on that frequency!

You can usually tell when a DX station is working split. If you hear a DX station making contacts and you don't hear the other sides of the conversations, suspect a split-frequency operation. As you tune above the DX station's frequency, you'll usually find a cacophony of desperate stations—the classic *pileup*.

Hunting long-distance (DX) contacts will bring rewards such as this QSL card from FOØAAA, the 2000 Clipperton Island DXpedition.

CLIPPERTON ISLAND 2000

FOØAAA

Low power (QRP) transceivers like these will get you on the air for less than $200.

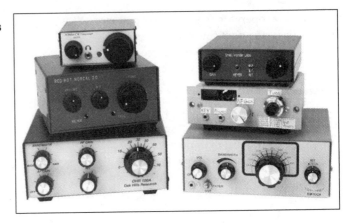

Experienced DX stations will often announce that they're working split. For example, on SSB you may hear the operator say, "Listening 14.210 to 14.220." On CW you might copy: UP 10, meaning that he's listening 10 kHz above his transmitting frequency.

The DXCC Award

One of the driving forces of the DX obsession is the coveted DX Century Club award, otherwise known as the DXCC. The DXCC award program is managed by the ARRL. The idea is relatively simple: Make contacts with stations in as many "official" DX countries as possible. You need 100 contacts—confirmed with QSL cards—to bag the initial DXCC certificate, and there are endorsement stickers for contacts you make beyond that point. At the top of the pile are the DX elite who've worked nearly every DXCC country in the world. They receive plaques and are admitted into the hallowed sanctum of the DXCC Honor Roll.

The definition of what makes an official DXCC country is quite different than what you might imagine. Did you know that Alaska and Hawaii are considered separate DXCC countries?

You can take your HF operating on the road by installing a mobile antenna on your car. This particular model is the Comet UHV-6 and it offers coverage from 40 through 2 meters.

Many DXCC "countries" are nothing more than tiny uninhabited islands. Groups of hams spend thousands of dollars (and even risk their lives) to mount *DXpeditions* on these godforsaken rocks. They operate for several days or several weeks, giving hams throughout the world the opportunity to confirm that they've worked one of these rare countries. (For more than you ever wanted to know about DXCC and DXing in general, pick up a copy of *The ARRL Operating Manual*.)

This certificate is one of the most sought-after pieces of paper in the amateur world: the DXCC award.

The downside of the global DXCC competition is that it's nearly impossible to enjoy a real conversation with a ham if he or she is in a rare or semi-rare country. You can have fascinating, long-winded chats with Britons, Germans, French, Japanese, Russians, Australians and so on. Hams in these and other countries are so numerous that they're never "rare." But don't expect a long exchange with a ham in, say, Ghana. As soon as the word gets around that he's on the air, the pileup begins. It's hard to have a friendly chat when other folks are demanding to be heard.

JUMPING INTO THE NETS

Whenever you have a group of hams who meet on a particular frequency at a particular time, you have a *net*. There are nets devoted to just about every purpose you can imagine. If you're a user of Kenwood equipment, there is a Kenwood user's net. If you're into amateur satellites, you can hang out with the AMSAT nets. DX nets are devoted to arranging contacts with difficult-to-work stations. *Traffic handling* is a popular net activity. Hams meet on various frequencies to relay messages from throughout the nation or the world. And then other nets are simply groups of hams who like to meet and talk about whatever is on their minds.

A net would dissolve into chaos if it wasn't for the *net control station*, or *NCS*. He or she acts like a traffic cop at a busy intersection. The NCS "calls" the net at the appropriate time. If you happen to tune in at the start of an SSB net, you might hear something like this . . .

"Attention all amateurs. Attention all amateurs. This is KX6Y, net-control for the Klingon Language Net. This net meets every Saturday at 1400 UTC on this frequency to exchange news and views concerning the language of the Klingon homeworld. Any stations wishing to join the net please call KX6Y now."

That's the cue for any interested ham to throw out his or her call sign. The NCS writes down each call sign in the order in which he receives it. At some point he may break in and say, "Okay, so far I have WB8IMY, N6ATQ, WR1B, KU7G and K1ZZ. Anyone else?"

Once the list is complete, the NCS will call each station in turn and ask if they have any questions or comments for the net. If he calls you, you have the option of speaking your piece or telling the NCS to skip to the next person. The NCS will also ask for new check-ins from time to time. If you didn't join the net at the beginning, that's your chance.

Traffic and other public-service nets are more tightly organized and follow stricter rules concerning who can say what . . . and when. For more information about these types of nets and how they work, see *The ARRL Operating Manual*.

CONTEST THE ISSUE

Contesting is *hot* on the HF bands. How hot? Consider the fact that large numbers of hams devote incredible amounts of time, and huge sums of money, in the pursuit of top contest scores. Their radio rooms are filled with computers running the latest contest software and heavy-duty RF amplifiers. Their property is a virtual forest of copper, steel and aluminum.

Although the rules differ, the goal of every Amateur Radio contest is to contact as many stations as possible. Through a combination of technology—and skills born of rugged experience—elite contesters consistently vault to the tops of the heaps. But contesting can be fun and worthwhile for even the smallest operators. All you have to do is set *realistic* goals (see the sidebar "The Casual Contester").

The name of the game in contesting is speed. You need to make the contact, grab the necessary information, and get on to the next contact *immediately*. This is especially true when you're *running*. Running is the practice of staying on one frequency and calling "CQ contest." If you have a large station and a large signal, running is a worthwhile strategy. It's also true if you are lucky enough to live in a highly desirable location—at least as far as the contest is concerned! For example, if you took a ham vacation to operate from the Yukon during Sweepstakes, you could sit and run stations throughout the entire contest. (You'd only need to change bands as propagation conditions shifted.) Your "rare" signal from the Yukon would be a magnet for Sweepstakes participants everywhere!

A contest exchange is usually quite short. As the hard-boiled cops in the old *Dragnet* TV series used to say, "Just the facts, madam." Check the contest rules in *QST* well ahead of time and make sure you understand the exchange format. The simplest exchange is a contact number followed by a signal report. The first contact you make is 001. Your second contact is 002 and so on. Signal reports are often given as 59 or 599—even when they aren't. Yes, this probably seems less than honest, but tossing out 59/599 reports has become the contesting custom. Here is how a typical contest exchange sounds on SSB:

Sander Idelson, KB1FPU (background) and Chris McCarthy, KB1ELV, stayed up all night to compete in the ARRL Phone Sweepstakes.

Table 7-1
Major Contests

Month	Contest	Exchange
Jan	ARRL VHF Sweepstakes	grid-square locator
Jan	CQ Worldwide 160-Meter Contest (CW)	signal report and state
Jan	ARRL RTTY Roundup	signal report and state
Feb	ARRL International DX Contest (CW)	signal report and state
Feb	CQ Worldwide 160-Meter Contest (phone)	see above
Mar	ARRL International DX Contest (phone)	see above
Mar	CQ WPX Contest (phone)	signal report and number
May	CQ WPX Contest (CW)	see above
Jun	ARRL June VHF QSO Party	grid-square locator
Jun	All Asian DX Contest (phone)	signal report and age
Jul	IARU HF World Championship	signal report and ITU zone
Aug	Worked All Europe (CW)	signal report and number
Sep	Worked All Europe (phone)	see above
Sep	ARRL September VHF QSO Party	grid-square locator
Oct	CQ Worldwide DX Contest (phone)	signal report and CQ zone
Nov	ARRL Sweepstakes (CW)	power level, serial number, year licensed
Nov	ARRL Sweepstakes (phone)	see above
Dec	ARRL 160-Meter Contest (CW)	signal report and section
Dec	ARRL 10-Meter Contest (phone and CW)	signal report and state

The Casual Contester

By Glenn Swanson, KB1GW

For many, participation in Amateur Radio contests is casual in the extreme. They hear the flurry of contest activity and decide to give out a few contacts. They're in, out and done. They savor a little of the contest thrill and that's sufficient. Others spend more time and energy, but they manage to keep their cools at the same time.

To the hams who chase countries (otherwise known as DXers), the worldwide contests are opportunities to look for contacts with rare countries. Many of these countries seem to be more "radio active" during major competitions, thus offering a DX hunter the chance to bag a rare one. For a number of DXers, the contest is simply a means to an end. They don't care if they earn big scores or not.

One way to find your own comfort level is to set a goal. Start with, say, making 25 contacts in a particular contest. The next time out, increase your goal (try making 50 contacts this time). Most of all, don't spend a single second worrying about how others are doing.

For the first few outings, just compete against yourself. Once you've gained proficiency (and some confidence), you can set out in the pursuit of other contesters. Set your sights on topping the score of one of the lower-scoring locals (check the post-contest score listings published in the various Amateur Radio magazines, such as *QST*). Copy some of the totals of the lower-scoring stations in your area and file them in a safe place. When the same contest runs again, post the scores at your operating position. They're your targets!

During one contest I used these scores as benchmarks to see how I was doing. Before I knew it, I found myself having a great time crossing off the scores as I surpassed them one by one. "Hey! I'm doing pretty good!" It took a while, but hour after hour I was able to gauge my progress. A wide smile came with each new line drawn through the next score on my list. Did I cross out all of them? Nope—not even close, but I had fun!

As an added bonus, my contest efforts have helped me to identify station improvements that I'd like to make. Now I have some new goals—like a faster computer and better antennas for 40 and 160 meters. As my abilities (and those of my station) increase, my confidence grows.

Enjoying the Team Approach

When I began contesting I did it solo—just me and my radio. Then I heard about something called a "multi-multi" contest station. By asking some of the contesters who hang out on the local repeater, I found out that this term was used to describe a contest station that had multiple-transmitters with multiple operators—all at a single location. That sounds like a big-time operation! It wasn't long before some of the local contesters got tired of my questions and invited me to see for myself. "How'd you like to come out and operate during the next contest?" Were they kidding? That was like asking a teenager if he'd like to drive a new Ferrari. I could hardly wait!

Of course, my next emotion was fear. What if I made a fool of myself? I'd never been at the controls of a big station, especially one manned by several grizzled contest veterans. As it turned out, I didn't need to worry. I found that these folks were very helpful. In fact, they were used to helping newcomers (like me) become acquainted with the operation of the multi-multi station. I also discovered the thrill of being a part of a team effort. We each had a hand in the operation of the station and in the resulting score. It was a pleasure to operate from a well-equipped station. There's nothing like playing with top-notch radios connected to large antenna arrays.

Believe it or not, you can operate from your own station and still be part of a team.

Some contests have special categories for "club competition." (The ARRL International DX Contest is one of these.) If you join a regional contest club, you can participate as part of your club's effort. For example, I belong to the Yankee Clipper Contest Club (YCCC), located here in New England. As a member I can participate in a contest from home and, after the contest, submit my score along with a notation that says it should apply to the YCCC's overall score. My score will be added to those submitted by other YCCC members who took part in the contest. Regional contest clubs encourage *all* of their members to submit a score. My score, no matter how small, will help the YCCC team.

Station Automation Makes it Easier

Thanks to the work of proficient software authors (who are also contesters), there are several contest-logging programs available. These programs take the place of a standard paper log by keeping track of all the details (such as time, date, and frequency) of each contact. As you enter the call sign of each station you contact, the program keeps a running tally of your contest score. They'll also check for duplicate contacts that may not count toward your score. This is casual contesting at its finest!

Many software packages (along with the appropriate hardware), are capable of *automating* your station. You can control many station operations via the keyboard of a personal computer. Contest programs have become so sophisticated that a competitive contester is handicapped without them. If your radio is set up for computer control, these programs can change things like the operating frequency and mode (including "split" operations) from the keyboard of your PC. Additional hardware options allow you to use the keyboard to send (preprogrammed) Morse-code messages, handle your antenna switching tasks and even point your antenna in the proper direction. The integration of these programs into a contest station has allowed many hams to operate their equipment with just a few keystrokes. To find pricing and ordering information for these software packages, try checking the advertisements found in *QST* and in the *National Contest Journal*. You'll also discover information about station automation and contesting in general in *The ARRL Operating Manual*.

Contest Speak

Call used: The call sign used by this op (operator) during the contest.

Claimed score: The total estimated score (not officially accepted yet).

Class: Entry classification. (Single operator, single operator with more than one transmitter, etc.) "Assisted" indicates that a spotting network such as a *DX PacketCluster* was also used.

Dupe: More than one contact with the same station. A duplicate contact may or may not count toward your final score, depending upon the rules of the contest. For example, the contest rules might allow dupes if they take place on two different bands.

Mults: Multipliers worked during the contest. A multiplier does what it says—it "multiplies" your point totals. Multipliers are stations located in specific states, zones, or countries according to the rules of the contest.

Qs: The total number of contacts made during the contest

Run: Working many stations, one after the other, on the same frequency.

Rate: The number of contacts per hour. Many contest programs will give you a numerical readout of your rate. Some will even show this information in graph form

Search and pounce: Searching the bands for the multiplier stations you need. Searching and pouncing is common when you have difficulty "running" stations on a particular frequency.

"CQ contest, CQ contest from WB8IMY."

"AB7AZ"

"AB7AZ from WB8IMY. You're 59, number 025." (He's my 25th contact.)

"QSL. You're also 59, number 100." (I'm his 100th contact.)

"Thanks, QRZ, WB8IMY."

Other exchanges may be more elaborate. You might have to give the last two digits of the year when you were first licensed, or your *CQ* or ITU zone number. (*The ARRL Operating Manual* contains complete *CQ* and ITU zone maps.)

Even if you never send in your logs for an official tally, you can still participate in a contest. A contact point is a contact point as far as the other contest stations are concerned. They'll accept your contacts and smile—even if they know that you won't show up in the scoring roster. The brief exchange counts toward *their* score, and that's all that matters.

Contests can also be quite valuable for the noncontester. An international DX contest will bring out all sorts of DX stations. Spend a couple of hours in this contest and you'll earn many contacts toward your DXCC award. In the same way, domestic contests are terrific for bagging your Worked All States award.

A contest is also a superb opportunity to test your antenna system. After the competition you can analyze your logs and determine which areas of the country (or the world) where your signal seems strongest (or weakest).

Info Guide

Unless otherwise indicated, all items in this section may be purchased, subject to availability, from your dealer or from ARRL—the national association for Amateur Radio, 225 Main St, Newington, CT 06111; tel 888-277-5289 (toll free); or on the Web at **www.arrl.org/catalog/**.

FM AND REPEATERS

Books and Software

The ARRL Handbook for Radio Amateurs covers the Amateur Radio state of the art—analog and digital electronics theory, antennas, transmitters, receivers, test equipment, other accessories and much more.

The ARRL Operating Manual is the most complete guide to Amateur Radio operating ever published. Chapter 3 is devoted to FM and repeaters.

The ARRL Repeater Directory, updated annually, is a pocket-sized listing of 20,000 repeaters in the US and Canada on 29, 50-54 and 144-148, 222-225, 420-450, 902-928 and 1240 MHz and above. Operating tips, band plan charts, frequency coordinator listings and more.

TravelPlus for Repeaters is CD-ROM software that will locate repeaters along the roadways you travel, anywhere in the US and Canada.

DIGITAL COMMUNICATION

Books

Practical Packet Radio by Stan Horzepa, WA1LOU, is a comprehensive treatment of packet communication. This books dissects the TNC and tells you how it works. You'll also learn the gritty details of packet nodes, networking and more.

Packet: Speed, More Speed and Applications is for packet enthusiasts interested in medium- to high-speed packet systems or applications that go beyond everyday messaging.

APRS Tracks, Maps and Mobiles by Stan Horzepa, WA1LOU, is the most comprehensive guide to the Automatic Position Reporting System in existence.

The ARRL HF Digital Handbook, by Steve Ford, WB8IMY, takes you on a tour through the worlds of RTTY, AMTOR, PACTOR, PSK31, G-TOR and CLOVER. You'll discover how to set up your station and communicate with each of these fascinating modes.

Equipment Manufacturers

Kantronics, 1202 East 23rd St, Lawrence, KS 66046-5006; tel 913-842-7745; **www.kantronics.com**.

MFJ Enterprises, Box 494, Mississippi State, MS 39762; tel 601-323-5869; **www.mfjenterprises.com**.

VHF/UHF SSB AND CW

Books

Beyond Line of Sight: A History of VHF Propagation from the Pages of QST explores the ways hams helped discover and exploit the propagation modes that allow VHF signals to travel hundreds and even thousands of miles. It's a subject all hams will find fascinating.

VHF/UHF Manual, from RSGB, is must reading for the VHF and UHF enthusiast. You'll find information on the history of VHF/UHF communications, propagation, tuned circuits, receivers, transmitters, integrated equipment, filters, antennas, microwaves, space communications, and test equipment.

Organizations

Central States VHF Society, Bruce Richardson, W9FZ, 2330 Lexington Ave S, #312, Mendota Heights, MN 55120, Web: **www.csvhfs.org/**

SMIRK—Six Meter International Radio Klub, PO Box 393, Junction, TX 76849, Web: **www.smirk.org/**

SWOT—Sidewinders on Two, c/o Howard Hallman, WD5DJT, 3230 Springfield, Lancaster, TX 75134-1214, Web: **home.swbell.net/wd5djt/**

The Six Meter Club, PO Box 307, Hatfield, AR 71945, Web: **6mt.com/club.htm**

UK Six Metre Group, **www.uksmg.org/**

SATELLITES

Books

The Radio Amateur's Satellite Handbook, written by Martin Davidoff, K2UBC, provides the ultimate reference for the satellite operator. All active satellites are covered in detail, including telemetry formats, uplink/downlink frequencies, on-board power systems and more.

ARRL Satellite Anthology is a collection of the best satellite articles from recent issues of *QST* and elsewhere. A must for every satellite operator.

Newsletters

The AMSAT Journal—available from AMSAT, PO Box 27, Washington, DC 20044; tel 301-589-6062; **www.amsat.org/**.

Software

Software for satellite tracking, telemetry decoding and PACSAT operation is available from: AMSAT, PO Box 27, Washington, DC 20044; tel 301-589-6062; **www.amsat.org/**.

Equipment Manufacturers

Down East Microwave, 954 Rte 519, Frenchtown, NJ 08825; tel 908-996-3584; **www.downeastmicrowave.com/** (downconverters, receive preamplifiers, antennas)

Hamtronics, 65-Q Moul Rd, Hilton, NY 14468; tel 716-392-9430; **www.hamtronics.com/** (downconverters)

SSB Electronic USA, 124 Cherrywood Dr, Mountaintop, PA 18707; tel 717-868-5643; **www.ssbusa.com/** (downconverters, receive preamplifiers, antennas)

AMATEUR TELEVISION

Equipment Manufacturers

Communication Concepts Inc, 508 Millstone Dr, Xenia, OH 45385; tel 513-426-8600; **www.communication-concepts.com/**.

PC Electronics, 2522 Paxson Ln, Arcadia, CA 91007-8537; tel 626-447-4565; **www.hamtv.com/**.

Supercircuits, 1403 Bayview Dr, Hermosa Beach, CA 90254; tel 800-335-9777; **www.supercircuits.com/**.

CONTACTING ARRL HEADQUARTERS VIA THE INTERNET

Awards	**awards@arrl.org**
Contests	**contests@arrl.org**
DXCC	**dxcc@arrl.org**
Educational Activities	**ead@arrl.org**
QEX magazine	**qex@arrl.org**
QST magazine	**qst@arrl.org**
Interference problems	**rfi@arrl.org**
Technical questions	**tis@arrl.org**
Exams	**vec@arrl.org**
Other inquiries	**hq@arrl.org**

ARRLWeb has a wealth of information for all hams, including latest news bulletins, an up-to-date hamfest calendar, technical information, publications catalog, links to other Amateur Radio-related WWW pages and much, much more! Just point your web browser to **www.arrl.org/**.

US Amateur Bands

April 15, 2000

160 METERS

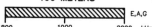

E,A,G

1800   1900   2000 kHz

Amateur stations operating at 1900–2000 kHz must not cause harmful interference to the radiolocation service and are afforded no protection from radiolocation operations.

80 METERS

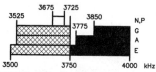

3675 3725
3525   3850   N,P
        3775   G
                        A
                        E
3500   3750   4000 kHz

5167.5 kHz (SSB only): Alaska emergency use only.

40 METERS

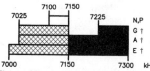

7100 7150
7025   7225   N,P
                        G †
                        A †
                        E †
7000   7150   7300 kHz

† Phone and image modes are permitted between 7075 and 7100 kHz for FCC licensed stations in ITU Regions 1 and 3 and by FCC licensed stations in ITU Region 2 West of 130 degrees West longitude or South of 20 degrees North latitude. See Sections 97.305(c) and 97.307(f)(11). Novice and Technician Plus licensees outside ITU Region 2 may use CW only between 7050 and 7075 kHz. See Section 97.301(e). These exemptions do not apply to stations in the continental US.

30 METERS

E,A,G

10,100   10,150 kHz

Maximum power on 30 meters is 200 watts PEP output. Amateurs must avoid interference to the fixed service outside the US.

20 METERS

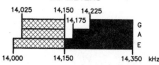

14,025   14,150 14,225
            14,175
                        G
                        A
                        E
14,000   14,150   14,350 kHz

17 METERS

E,A,G

18,068   18,110   18,168 kHz

15 METERS

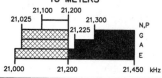

21,100 21,200
21,025   21,300   N,P
        21,225   G
                        A
                        E
21,000   21,200   21,450 kHz

Novice, Advanced and Technician Plus Allocations

New Novice, Advanced and Technician Plus licenses have not been issued since April 15, 2000. However, the FCC has allowed the frequency allocations for these license classes to remain in effect. They will continue to renew existing licenses for those classes.

12 METERS

E,A,G

24,890   24,930   24,990 kHz

10 METERS

28,100 28,500
                        N,P
28,000 28,300   E,A,G
                        29,700 kHz

Novices and Technician Plus licensees are limited to 200 watts PEP output on 10 meters.

6 METERS

50.1   E,A,G,P,T
50.0   54.0 MHz

2 METERS

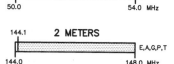

144.1   E,A,G,P,T
144.0   148.0 MHz

1.25 METERS

E,A,G,P,T,N

222.0   225.0 MHz

Novices are limited to 25 watts PEP output from 222 to 225 MHz.

70 CENTIMETERS **

E,A,G,P,T

420.0   450.0 MHz

33 CENTIMETERS **

E,A,G,P,T

902.0   928.0 MHz

23 CENTIMETERS **

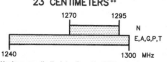

1270   1295
                        N
                        E,A,G,P,T
1240   1300 MHz

Novices are limited to 5 watts PEP output from 1270 to 1295 MHz.

US AMATEUR POWER LIMITS

At all times, transmitter power should be kept down to that necessary to carry out the desired communications. Power is rated in watts PEP output. Unless otherwise stated, the maximum power output is 1500 W. Power for all license classes is limited to 200 W in the 10,100–10,150 kHz band and in all Novice subbands below 28,100 kHz. Novices and Technicians with Morse code credit are restricted to 200 W in the 28,100–28,500 kHz subbands. In addition, Novices are restricted to 25 W in the 222–225 MHz band and 5 W in the 1270–1295 MHz subband.

Operators with Technician class licenses and above may operate on all bands above 50 MHz. For more detailed information see The FCC Rule Book.

KEY

= CW, RTTY and data

= CW, RTTY, data, MCW, test, phone and image

= CW, phone and image

= CW and SSB phone

= CW, RTTY, data, phone, and image

= CW only

E = EXTRA CLASS
A = ADVANCED
G = GENERAL
P = TECHNICIAN PLUS
T = TECHNICIAN
N = NOVICE

* Technicians passing the Morse code exam will gain HF Novice privileges, although they still hold a Technician license.

** Geographical and power restrictions apply to these bands. See The FCC Rule Book for more information about your area.

Above 23 Centimeters:

All licensees except Novices are authorized all modes on the following frequencies:
2300–2310 MHz
2390–2450 MHz
3300–3500 MHz
5650–5925 MHz
10.0–10.5 GHz
24.0–24.25 GHz
47.0–47.2 GHz
75.5–81.0 GHz
119.98–120.02 GHz
142–149 GHz
241–250 GHz
All above 300 GHz

For band plans and sharing arrangements, see the FCC Rule Book.

SOME ABBREVIATIONS FOR CW WORK

Some Abbreviations help to cut down unnecessary transmission; do not abbreviate unnecessarily when working an operator of unknown experience.

AA	All after
AB	All before
ABT	About
ADR	Address
AGN	Again
ANT	Antenna
BCI	Broadcast Interference
BCL	Broadcast listener
BK	Break, break me; break-in
BN	All between; been
BUG	Semi-automatic key
B4	Before
C	Yes
CFM	Confirm; I confirm
CK	Check
CL	I am closing my station; call
CLD-CLG	Called; calling
CQ	Calling any station
CUD	Could
CUL	See you later
CW	Continuous wave (i.e., radiotele-graph)
DLD-DLVD	Delivered
DR	Dear
DX	Distance, foreign countries
ES	And,
FB	Fine business, excellent
FM	Frequency modulation
GA	Go ahead (or resume sending)
GB	Good-by
GBA	Give better address
GE	Good evening
GG	Going
GM	Good morning
GN	Good night
GND	Ground
GUD	Good
HI	The telegraphic laugh
HR	Here, hear
HV	Have
HW	How
LID	A poor operator
MA, MILS	Milliamperes
MSG	Message; prefix to radiogram
N	No
NCS	Net control station
ND	Nothing doing
NIL	Nothing; I have nothing for you
NM	No more
NR	Number
NW	Now; I resume transmission
OB	Old boy

OC	Old chap
OM	Old man
OP-OPR	Operator
OT	Old timer, old top
PBL	Preamble
PSE	Please
PWR	Power
PX	Press
R	Received as transmitted; are
RCD	Received
RCVR (RX)	Receiver
REF	Refer to; referring to; reference
RFI	Radio frequency interference
RIG	Station equipment
RPT	Repeat; I repeat; report
RTTY	Radioteletype
RX	Receiver
SASE	Self-addressed, stamped envelope
SED	Said
SIG	Signature; signal
SINE	Operator's personal initials or nickname
SKED	Schedule
SRI	Sorry
SSB	Single sideband
SVC	Service; prefix to service message
T	Zero
TFC	Traffic
TMW	Tomorrow
TNX-TKS	Thanks
TT	That
TU	Thank you
TVI	Television interference
TX	Transmitter
TXT	Text
UR-URS	Your; you're; yours
VFO	Variable-frequency oscillator
VY	Very
WA	Word after
WB	Word before
WD-WDS	Word; words
WKD-WKG	Worked; working
WL	Well; will
WUD	Would
WX	Weather
XCVR	Transceiver
XMTR (TX)	Transmitter
XTAL	Crystal
XYL (YF)	Wife
YL	Young lady
73	Best regards
88	Love and kisses

STANDARD ITU PHONETICS

A—Alfa (**AL** FAH)
B—Bravo (**BRAH** VOH)
C—Charlie (**CHAR** LEE) or (**SHAR** LEE)
D—Delta (**DELL** TAH)
E—Echo (**ECK** OH)
F—Foxtrot (**FOKS** TROT)
G—Golf (GOLF)
H—Hotel (HOH **TELL**)
I—India (**IN** DEE AH)
J—Juliett (**JEW** LEE ETT)
K—Kilo (**KEY** LOH)
L—Lima (**LEE** MAH)
M—Mike (MIKE)

N—November (NO **VEM** BER)
O—Oscar (**OSS** CAH)
P—Papa (PAH **PAH**)
Q—Quebec (KEH **BECK**)
R—Romeo (**ROW** ME OH)
S—Sierra (SEE **AIR** RAH)
T—Tango (**TANG** GO)
U—Uniform (**YOU** NEE FORM) or (OO NEE FORM)
V—Victor (**VIK** TAH)
W—Whiskey (**WISS** KEY)
X—X-RAY (**ECKS** RAY)
Y—Yankee (**YANG** KEY)
Z—Zulu (**ZOO** LOO)

Note: The **boldfaced** syllables are emphasized. The pronunciations shown in this table were designed for those who speak any of the international languages. The pronunciations given for "Oscar" and "Victor" may seem awkward to English-speaking people in the US.

Q SIGNALS

These Q signals are the ones used most often on the air. (Q abbreviations take the form of questions only when they are sent followed by a question mark.)

QRG	Will you tell me my exact frequency (or that of ___)? Your exact frequency (or that of ___) is ___ kHz.
QRL	Are you busy? I am busy (or I am busy with ___). Please do not interfere.
QRM	Is my transmission being interfered with? Your transmission is being interfered with ___ (1. Nil; 2. Slightly; 3. Moderately; 4. Severely; 5. Extremely.)
QRN	Are you troubled by static? I am troubled by static ___. (1-5 as under QRM.)
QRO	Shall I increase power? Increase power.
QRP	Shall I decrease power? Decrease power.
QRQ	Shall I send faster? Send faster (___ WPM).
QRS	Shall I send more slowly? Send more slowly (___ WPM).
QRT	Shall I stop sending? Stop sending.
QRU	Have you anything for me? I have nothing for you.
QRV	Are you ready? I am ready.
QRX	When will you call me again? I will call you again at ___ hours (on ___ kHz).
QRZ	Who is calling me? You are being called by ___(on ___kHz).
QSB	Are my signals fading? Your signals are fading.
QSK	Can you hear me between your signals and if so can I break in on your transmission? I can hear you between signals; break in on my transmission.
QSL	Can you acknowledge receipt (of a message or transmission)? I am acknowledging receipt.
QSN	Did you hear me (or ___) on ___kHz? I did hear you (or ___) on ___kHz.
QSO	Can you communicate with ___direct or by relay? I can communicate with ___ direct (or relay through ___).
QSP	Will you relay to ___? I will relay to ___.
QST	General call preceding a message addressed to all amateurs and ARRL members. This is in effect "CQ ARRL."
QSX	Will you listen to ___on ___kHz? I am listening to _____ on ___kHz.
QSY	Shall I change to transmission on another frequency? Change to transmission on another frequency (or on ___ kHz).
QTB	Do you agree with my counting of words? I do not agree with your counting of words. I will repeat the first letter or digit of each word or group.
QTC	How many messages have you to send? I have ___ messages for you (or for ___).
QTH	What is your location? My location is ___.
QTR	What is the correct time? The time is ___.

W1AW's schedule is at the same local time throughout the year. The schedule according to your local time will change if your local time does not have seasonal adjustments that are made at the same time as North American time changes between standard time and daylight time. From the first Sunday in April to the last Sunday in October, UTC = Eastern Time + 4 hours. For the rest of the year, UTC = Eastern Time + 5 hours.

W1AW SCHEDULE

Pacific	Mtn	Cent	East	Mon	Tue	Wed	Thu	Fri
6 AM	7 AM	8 AM	9 AM		Fast Code	Slow Code	Fast Code	Slow Code
7 AM-1 PM	8 AM-2 PM	9 AM-3 PM	10 AM-4 PM	Visiting Operator Time (12 PM - 1 PM closed for lunch)				
1 PM	2 PM	3 PM	4 PM	Fast Code	Slow Code	Fast Code	Slow Code	Fast Code
2 PM	3 PM	4 PM	5 PM	Code Bulletin				
3 PM	4 PM	5 PM	6 PM	Teleprinter Bulletin				
4 PM	5 PM	6 PM	7 PM	Slow Code	Fast Code	Slow Code	Fast Code	Slow Code
5 PM	6 PM	7 PM	8 PM	Code Bulletin				
6 PM	7 PM	8 PM	9 PM	Teleprinter Bulletin				
6⁴⁵ PM	7⁴⁵ PM	8⁴⁵ PM	9⁴⁵ PM	Voice Bulletin				
7 PM	8 PM	9 PM	10 PM	Fast Code	Slow Code	Fast Code	Slow Code	Fast Code
8 PM	9 PM	10 PM	11 PM	Code Bulletin				

• MORSE CODE TRANSMISSIONS:

Frequencies are 1.818, 3.5815, 7.0475, 14.0475, 18.0975, 21.0675, 28.0675 and 147.555 MHz.

Slow Code = practice sent at 5, 7½, 10, 13 and 15 wpm.

Fast Code = practice sent at 35, 30, 25, 20, 15, 13 and 10 wpm.

Code practice text is from the pages of *QST*. The source is given at the beginning of each practice session and alternate speeds within each session. For example, "Text is from July 2001 *QST*, pages 9 and 81," indicates that the plain text is from the article on page 9 and mixed number/letter groups are from page 81.

Code bulletins are sent at 18 wpm.

W1AW qualifying runs are sent on the same frequencies as the Morse code transmissions. West Coast qualifying runs are transmitted on approximately 3.590 MHz by K6YR. At the beginning of each code practice session, the schedule for the next qualifying run is presented. Underline one minute of the highest speed you copied, certify that your copy was made without aid, and send it to ARRL for grading. Please include your name, call sign (if any) and complete mailing address. Send a 9×12-inch SASE for a certificate, or a business-size SASE for an endorsement.

• TELEPRINTER TRANSMISSIONS:

Frequencies are 3.625, 7.095, 14.095, 18.1025, 21.095, 28.095 and 147.555 MHz.

Bulletins are sent at 45.45-baud Baudot and 100-baud AMTOR, FEC Mode B. 110-baud ASCII will be sent only as time allows.

On Tuesdays and Fridays at 6:30 PM Eastern Time, Keplerian elements for many amateur satellites are sent on the regular teleprinter frequencies.

• VOICE TRANSMISSIONS:

Frequencies are 1.855, 3.99, 7.29, 14.29, 18.16, 21.39, 28.59 and 147.555 MHz.

• MISCELLANEA:

On Fridays, UTC, a DX bulletin replaces the regular bulletins.

W1AW is open to visitors from 10 AM until noon and from 1 PM until 3:45 PM on Monday through Friday. FCC licensed amateurs may operate the station during that time. Be sure to bring your current FCC amateur license or a photocopy.

In a communication emergency, monitor W1AW for special bulletins as follows: voice on the hour, teleprinter at 15 minutes past the hour, and CW on the half hour.

Headquarters and W1AW are closed on New Year's Day, President's Day, Good Friday, Memorial Day, Independence Day, Labor Day, Thanksgiving and the following Friday, and Christmas Day.

THE "CONSIDERATE OPERATOR'S FREQUENCY GUIDE"

The following frequencies are generally recognized for certain modes or activities (all frequencies are in MHz).

Nothing in the rules recognizes a net's, group's or any individual's special privilege to any specific frequency. Section 97.101(b) of the Rules states that "Each station licensee and each control operator must cooperate in selecting transmitting channels and in making the most effective use of the amateur service frequencies. No frequency will be assigned for the exclusive use of any station." No one "owns" a frequency.

It's good practice—and plain old common sense—for any operator, regardless of mode, to check to see if the frequency is in use prior to engaging operation. If you are there first, other operators should make an effort to protect you from interference to the extent possible, given that 100% interference-free operation is an unrealistic expectation in today's congested bands.

Frequency	Mode/Activity
1.800-1.830	CW, data and other narrowband modes
1.810	QRP CW calling frequency
1.830-1.840	CW, data and other narrowband modes, intercontinental QSOs only
1.840-1.850	CW; SSB, SSTV and other wideband modes, intercontinental QSOs only
1.850-2.000	CW; phone, SSTV and other wideband modes
3.500-3.510	CW DX
3.560	QRP
3.590	RTTY DX
3.580-3.620	Data
3.620-3.635	Automatically controlled data stations
3.710	QRP Novice/Technician CW calling frequency
3.790-3.800	DX window
3.845	SSTV
3.885	AM calling frequency
3.985	QRP SSB calling frequency
7.040	RTTY DX
	QRP CW calling frequency
7.075-7.100	Phone in KH/KL/KP only
7.080-7.100	Data
7.100-7.105	Automatically controlled data stations
7.171	SSTV
7.285	QRP SSB calling frequency
7.290	AM calling frequency
10.106	QRP CW calling frequency
10.130-10.140	Data
10.140-10.150	Automatically controlled data stations
14.060	QRP CW calling frequency
14.070-14.095	Data
14.095-14.0995	Automatically controlled data stations
14.100	IBP/NCDXF beacons
14.1005-14.112	Automatically controlled data stations
14.230	SSTV
14.285	QRP SSB calling frequency
14.286	AM calling frequency
18.100-18.105	Data
18.105-18.110	Automatically controlled data stations
21.060	QRP CW calling frequency
21.070-21.100	Data
21.090-21.100	Automatically controlled data stations
21.340	SSTV
21.385	QRP SSB calling frequency
24.920-24.925	Data
24.925-24.930	Automatically controlled data stations
28.060	QRP CW calling frequency
28.070-28.120	Data
28.120-28.189	Automatically controlled data stations
28.190-28.225	Beacons
28.385	QRP SSB calling frequency
28.680	SSTV
29.000-29.200	AM
29.300-29.510	Satellite downlinks
29.520-29.580	Repeater inputs
29.600	FM simplex
29.620-29.680	Repeater outputs

Note

ARRL band plans for frequencies above 28.300 MHz are shown in *The ARRL Repeater Directory* and *The FCC Rule Book*.

IBP/NCDXF beacons operate on 14.100, 18.110, 21.150, 24.930 and 28.200 MHz.

ABBREVIATIONS LIST

A

a—atto (prefix for 10^{-18})
A—ampere (unit of electrical current)
ac—alternating current
ACC—Affiliated Club Coordinator
ACSSB—amplitude-compandored single
  sideband
A/D—analog-to-digital
ADC—analog-to-digital converter
AF—audio frequency
AFC—automatic frequency control
AFSK—audio frequency-shift keying
AGC—automatic gain control
A/h—ampere hour
ALC—automatic level control
AM—amplitude modulation
AMRAD—Amateur Radio Research and
  Development Corp
AMSAT—Radio Amateur Satellite Corp
AMTOR—Amateur Teleprinting Over Radio
ANT—antenna
ARA—Amateur Radio Association
ARC—Amateur Radio Club
ARES—Amateur Radio Emergency Service
ARQ—Automatic repeat request
ARRL—American Radio Relay League
ARS—Amateur Radio Society (station)
ASCII—American National Standard Code for
  Information Interchange
ATV—amateur television
AVC—automatic volume control
AWG—American wire gauge
az-el—azimuth-elevation

B

B—bel; blower; susceptance; flux density
  (inductors)
balun—balanced to unbalanced (transformer)
BC—broadcast
BCD—binary coded decimal
BCI—broadcast interference
Bd—baud (bit/s in single-channel binary data
  transmission)
BER—bit error rate
BFO—beat-frequency oscillator
bit—binary digit
bit/s—bits per second
BM—Bulletin Manager
BPF—band-pass filter
BPL—Brass Pounders League

BT—battery
BW—bandwidth

C

c—centi (prefix for 10^{-2})
C—coulomb (quantity of electric charge);
  capacitor
CAC—Contest Advisory Committee
CATVI—cable television interference
CB—Citizens Band (radio)
CBBS—computer bulletin-board service
CBMS—computer-based message system
CCIR—See ITU-T
CCITT—See ITU-T
CCTV—closed-circuit television
CCW—coherent CW
ccw—counterclockwise
CD—civil defense
cm—centimeter
CMOS—complimentary-symmetry metal-oxide
  semiconductor
coax—coaxial cable
COR—carrier-operated relay
CP—code proficiency (award)
CPU—central processing unit
CRT—cathode ray tube
CT—center tap
CTCSS—continuous tone-coded squelch
  system
cw—clockwise
CW—continuous wave

D

d—deci (prefix for 10^{-1})
D—diode
da—deca (prefix for 10)
D/A—digital-to-analog
DAC—digital-to-analog converter
dB—decibel (0.1 bel)
dBi—decibels above (or below) isotropic
  antenna
dBm—decibels above (or below) one milliwatt
DBM—doubly balanced mixer
dBV—decibels above/below 1 V (in video,
  relative to 1 V P-P)
dBW—decibels above/below 1 W
dc—direct current
D-C—direct conversion
DDS—direct digital synthesis
DEC—District Emergency Coordinator
deg—degree

10

DET—detector
DF—direction finding; direction finder
DIP—dual in-line package
DMM—digital multimeter
DPDT—double-pole double-throw (switch)
DPSK—differential phase-shift keying
DPST—double-pole single-throw (switch)
DS—direct sequence (spread spectrum); display
DSB—double sideband
DSP—digital signal processing
DTMF—dual-tone multifrequency
DVM—digital voltmeter
DX—long distance; duplex
DXAC—DX Advisory Committee
DXCC—DX Century Club

E

e—base of natural logarithms (2.71828)
E—Voltage
EA—Educational Advisor
EC—Emergency Coordinator
ECL—emitter-coupled logic
EHF—extremely high frequency (30-300 GHz)
EIA—Electronic Industries Alliance
EIRP—effective isotropic radiated power
ELF—extremely low frequency
ELT—emergency locator transmitter
EMC—electromagnetic compatibility
EME—earth-moon-earth (moonbounce)
EMF—electromotive force
EMI—electromagnetic interference
EMP—electromagnetic pulse
EOC—emergency operations center
EPROM—erasable programmable read only memory

F

f—femto (prefix for 10^{-15}); frequency
F—farad (capacitance unit); fuse
fax—facsimile
FCC—Federal Communications Commission
FD—Field Day
FEMA—Federal Emergency Management Agency
FET—field-effect transistor
FFT—fast Fourier transform
FL—filter
FM—frequency modulation
FMTV—frequency-modulated television
FSK—frequency-shift keying
FSTV—fast-scan (real-time) television
ft—foot (unit of length)

G

g—gram (unit of mass)
G—giga (prefix for 10^9); conductance
GaAs—gallium arsenide
GDO—grid- or gate-dip oscillator
GHz—gigahertz (10^{9} ^{Hz})
GND—ground

H

h—hecto (prefix for 10^2)
H—henry (unit of inductance)
HF—high frequency (3-30 MHz)
HFO—high-frequency oscillator; heterodyne frequency oscillator
HPF—highest probable frequency; high-pass filter
Hz—hertz (unit of frequency, 1 cycle/s)

I

I—current, indicating lamp
IARU—International Amateur Radio Union
IC—integrated circuit
ID—identification; inside diameter
IEEE—Institute of Electrical and Electronics Engineers
IF—intermediate frequency
IMD—intermodulation distortion
in.—inch (unit of length)
in./s—inch per second (unit of velocity)
I/O—input/output
IRC—international reply coupon
ISB—independent sideband
ITU—International Telecommunication Union
ITU-T—Telecommunication Standardization Bureau of the ITU (combines activities of predecessor organizations CCIR and CCITT)

J

j—operator for complex notation, as for reactive component of an impedance ($+j$ inductive; $-j$ capacitive)
J—joule (kg m^2/s^2) (energy or work unit); jack
JFET—junction field-effect transistor

K

k—kilo (prefix for 10^3); Boltzmann's constant (1.38×10^{-23} J/K)
K—kelvin (used without degree symbol) (absolute temperature scale); relay
kBd—1000 bauds
kbit—1024 bits
kbit/s—1024 bits per second

kbyte—1024 bytes
kg—kilogram
kHz—kilohertz
km—kilometer
kV—kilovolt
kW—kilowatt
kΩ—kilohm

L

l—liter (liquid volume)
L—lambert; inductor
lb—pound (force unit)
LC—inductance-capacitance
LCD—liquid crystal display
LED—light-emitting diode
LF—low frequency (30-300 kHz)
LHC—left-hand circular (polarization)
LO—local oscillator; League Official
LP—log periodic
LS—loudspeaker
lsb—least significant bit
LSB—lower sideband
LSI—large-scale integration
LUF—lowest usable frequency

M

m—meter (length); milli (prefix for 10^{-3})
M—mega (prefix for 10^6); meter (instrument)
mA—milliampere
mAh—milliamperehour
MCP—multimode communications processor
MDS—Multipoint Distribution Service; minimum discernible (or detectable) signal
MF—medium frequency (300-3000 kHz)
mH—millihenry
MHz—megahertz
mi—mile, statute (unit of length)
mi/h—mile per hour
mi/s—mile per second
mic—microphone
min—minute (time)
MIX—mixer
mm—millimeter
MOD—modulator
modem—modulator/demodulator
MOS—metal-oxide semiconductor
MOSFET—metal-oxide semiconductor field-effect transistor
MS—meteor scatter
ms—millisecond
m/s—meters per second
msb—most-significant bit
MSI—medium-scale integration

MSK—minimum-shift keying
MSO—message storage operation
MUF—maximum usable frequency
mV—millivolt
mW—milliwatt
MΩ—megohm

N

n—nano (prefix for 10^{-9}); number of turns (inductors)
NBFM—narrow-band frequency modulation
NC—no connection; normally closed
NCS—net-control station; National Communications System
nF—nanofarad
NF—noise figure
nH—nanohenry
NiCd—nickel cadmium
NM—Net Manager
NMOS—N-channel metal-oxide silicon
NO—normally open
NPN—negative-positive-negative (transistor)
NPRM—Notice of Proposed Rule Making (FCC)
ns—nanosecond
NTS—National Traffic System

O

OBS—Official Bulletin Station
OD—outside diameter
OES—Official Emergency Station
OO—Official Observer
op amp—operational amplifier
ORS—Official Relay Station
OSC—oscillator
OSCAR—Orbiting Satellite Carrying Amateur Radio
OTC—Old Timer's Club
oz—ounce (force unit, 1/16 pound)

P

p—pico (prefix for 10^{-12})
P—power; plug
PA—power amplifier
PacTOR—digital mode combining aspects of packet and AMTOR
PAM—pulse-amplitude modulation
PC—printed circuit
P_D—power dissipation
PEP—peak envelope power
PEV—peak envelope voltage
pF—picofarad
pH—picohenry
PIA—Public Information Assistant

12

PIN—positive-intrinsic-negative (semi-conductor)
PIO—Public Information Officer
PIV—peak inverse voltage
PLL—phase-locked loop
PM—phase modulation
PMOS—P-channel (metal-oxide semiconductor)
PNP—positive-negative positive (transistor)
pot—potentiometer
P-P—peak to peak
ppd—postpaid
PROM—programmable read-only memory
PSHR—Public Service Honor Roll
PTO—permeability-tuned oscillator
PTT—push to talk

Q

Q—figure of merit (tuned circuit); transistor
QRP—low power (less than 5-W output)

R

R—resistor
RACES—Radio Amateur Civil Emergency Service
RAM—random-access memory
RC—resistance-capacitance
R/C—radio control
RCC—Rag Chewer's Club
RDF—radio direction finding
RF—radio frequency
RFC—radio-frequency choke
RFI—radio-frequency interference
RHC—right-hand circular (polarization)
RIT—receiver incremental tuning
RLC—resistance-inductance-capacitance
RM—rule making (number assigned to petition)
r/min—revolutions per minute
RMS—root mean square
ROM—read-only memory
r/s—revolutions per second
RST—readability-strength-tone (CW signal report)
RTTY—radioteletype
RX—receiver, receiving

S

s—second (time)
S—siemens (unit of conductance; switch
SASE—self-addressed stamped envelope
SCF—switched capacitor filter
SCR—silicon controlled rectifier
SEC—Section Emergency Coordinator
SET—Simulated Emergency Test

SGL—State Government Liaison
SHF—super-high frequency (3-30 GHz)
SM—Section Manager; silver mica (capacitor)
S/N—signal-to-noise ratio
SPDT—single pole double-throw (switch)
SPST—single-pole single-throw (switch)
SS—Sweepstakes; spread spectrum
SSB—single sideband
SSC—Special Service Club
SSI—small-scale integration
SSTV—slow-scan television
STM—Section Traffic Manager
SX—simplex
sync—synchronous, synchronizing
SWL—shortwave listener
SWR—standing-wave ratio

T

T—tera (prefix for 10^{12}); transformer
TA—Technical Advisor
TC—Technical Coordinator
TCC—Transcontinental Corps (NTS)
TCP/IP—Transmission Control Protocol/Internet Protocol
tfc—traffic
TNC—terminal node controller (packet radio)
TR—transmit/receive
TS—Technical Specialist
TTL—transistor-transistor logic
TTY—teletypewriter
TU—terminal unit
TV—television
TVI—television interference
TX—transmitter, transmitting

U

U—integrated circuit
UHF—ultra-high frequency (300 MHz to 3 GHz)
USB—upper sideband
UTC—Coordinated Universal Time
UV—ultraviolet

V

V—volt; vacuum tube
VCO—voltage-controlled oscillator
VCR—video cassette recorder
VDT—video-display terminal
VE—Volunteer Examiner
VEC—Volunteer Examiner Coordinator
VFO—variable-frequency oscillator
VHF—very-high frequency (30-300 MHz)

VLF—very-low frequency (3-30 kHz)
VLSI—very-large-scale integration
VMOS—V-topology metal-oxide semiconductor
VOM—volt-ohm meter
VOX—voice operated switch
VR—voltage regulator
VSWR—voltage standing-wave ratio
VTVM—vacuum-tube voltmeter
VUAC—VHF/UHF Advisory Committee
VUCC—VHF/UHF Century Club
VXO—variable-frequency crystal oscillator

W

W—watt (kg m^2s^{-3}), unit of power
WAC—Worked All Continents
WAS—Worked All States
WBFM—wide-band frequency modulation
WEFAX—weather facsimile
Wh—watthour
WPM—words per minute
WRC—World Radio Conference
WVDC—working voltage, direct current

X

X—reactance
XCVR—transceiver
XFMR—transformer
XIT—transmitter incremental tuning
XO—crystal oscillator
XTAL—crystal
XVTR—transverter

Y

Y—crystal; admittance
YIG—yttrium iron garnet

Z

Z—impedance; also see UTC

5BDXCC—Five-Band DXCC
5BWAC—Five-Band WAC
5BWAS—Five-Band WAS
6BWAC—Six-Band WAC

°—degree (plane angle)
°C—degree Celsius (temperature)
°F—degree Fahrenheit (temperature)
α—(alpha) angles; coefficients, attenuation constant, absorption factor, area, common-base forward current-transfer ratio of a bipolar transistor
β—(beta) angles; coefficients, phase constant current gain of common-emitter transistor amplifiers
γ—(gamma) specific gravity, angles, electrical conductivity, propagation constant
Υ—(gamma) complex propagation constant
δ—(delta) increment or decrement; density; angles
Δ—(delta) increment or decrement determinant, permittivity
ε—(epsilon) dielectric constant; permittivity; electric intensity
ζ—(zeta) coordinates; coefficients
η—(eta) intrinsic impedance; efficiency; surface charge density; hysteresis; coordinate
θ—(theta) angular phase displacement; time constant; reluctance; angles
ι—(iota) unit vector
κ—(kappa) susceptibility; coupling coefficient
λ—(lambda) wavelength; attenuation constant
Λ—(lambda) permeance
μ—(mu) permeability; amplification factor; micro (prefix for 10^{-6})
μC—microcomputer
μF—microfarad
μH—microhenry
μP—microprocessor
ξ—(xi) coordinates
π—(pi) 3.14159
ρ—(rho) resistivity; volume charge density; coordinates; reflection coefficient
σ—(sigma) surface charge density; complex propagation constant; electrical conductivity; leakage coefficient; deviation
Σ—(sigma) summation
τ—(tau) time constant; volume resistivity; time-phase displacement; transmission factor; density
φ—(phi) magnetic flux; angles
Φ—(phi) summation
χ—(chi) electric susceptibility; angles
ψ—(psi) dielectric flux; phase difference; coordinates; angles
ω—(omega) angular velocity $2\pi f$
Ω—(omega) resistance in ohms; solid angle

ALLOCATION OF INTERNATIONAL CALL SIGNS

Call Sign Series	Allocated to	Call Sign Series	Allocated to	Call Sign Series	Allocated to
AAA-ALZ	United States of America	EXA-EXZ	Kyrgyzstan	LAA-LNZ	Norway
AMA-AOZ	Spain	EYA-EYZ	Tajikistan	LOA-LWZ	Argentina
APA-ASZ	Pakistan	EZA-EZZ	Turkmenistan	LXA-LXZ	Luxembourg
ATA-AWZ	India	E2A-E2Z	Thailand	LYA-LYZ	Lithuania
AXA-AXZ	Australia	E3A-E3Z	Eritrea	LZA-LZZ	Bulgaria
AYA-AZZ	Argentina	†E4A-E4Z	Palestine	L2A-L9Z	Argentina
A2A-A2Z	Botswana	FAA-FZZ	France	MAA-MZZ	United Kingdom of Great Britain and Northern Ireland
A3A-A3Z	Tonga	GAA-GZZ	United Kingdom of Great Britain and Northern Ireland		
A4A-A4Z	Oman				
A5A-A5Z	Bhutan			NAA-NZZ	United States of America
A6A-A6Z	United Arab Emirates	HAA-HAZ	Hungary		
A7A-A7Z	Qatar	HBA-HBZ	Switzerland	OAA-OCZ	Peru
A8A-A8Z	Liberia	HCA-HDZ	Ecuador	ODA-ODZ	Lebanon
A9A-A9Z	Bahrain	HEA-HEZ	Switzerland	OEA-OEZ	Austria
BAA-BZZ	China	HFA-HFZ	Poland	OFA-OJZ	Finland
CAA-CEZ	Chile	HGA-HGZ	Hungary	OKA-OLZ	Czech Republic
CFA-CKZ	Canada	HHA-HHZ	Haiti	OMA-OMZ	Slovak Republic
CLA-CMZ	Cuba	HIA-HIZ	Dominican Republic	ONA-OTZ	Belgium
CNA-CNZ	Morocco			OUA-OZZ	Denmark
COA-COZ	Cuba	HJA-HKZ	Colombia	PAA-PIZ	Netherlands
CPA-CPZ	Bolivia	HLA-HLZ	South Korea	PJA-PJZ	Netherlands Antilles
CQA-CUZ	Portugal	HMA-HMZ	North Korea		
CVA-CXZ	Uruguay	HNA-HNZ	Iraq	PKA-POZ	Indonesia
CYA-CZZ	Canada	HOA-HPZ	Panama	PPA-PYZ	Brazil
C2A-C2Z	Nauru	HQA-HRZ	Honduras	PZA-PZZ	Suriname
C3A-C3Z	Andorra	HSA-HSZ	Thailand	P2A-P2Z	Papua New Guinea
C4A-C4Z	Cyprus	HTA-HTZ	Nicaragua		
C5A-C5Z	Gambia	HUA-HUZ	El Salvador	P3A-P3Z	Cyprus
C6A-C6Z	Bahamas	HVA-HVZ	Vatican City	P4A-P4Z	Aruba
*C7A-C7Z	World Meteorological Organization	HWA-HYZ	France	P5A-P9Z	North Korea
		HZA-HZZ	Saudi Arabia	RAA-RZZ	Russian Federation
		H2A-H2Z	Cyprus		
C8A-C9Z	Mozambique	H3A-H3Z	Panama	SAA-SMZ	Sweden
DAA-DRZ	Germany	H4A-H4Z	Solomon Islands	SNA-SRZ	Poland
DSA-DTZ	South Korea	H6A-H7Z	Nicaragua	•SSA-SSM	Egypt
DUA-DZZ	Philippines	H8A-H9Z	Panama	•SSN-STZ	Sudan
D2A-D3Z	Angola	IAA-IZZ	Italy	SUA-SUZ	Egypt
D4A-D4Z	Cape Verde	JAA-JSZ	Japan	SVA-SZZ	Greece
D5A-D5Z	Liberia	JTA-JVZ	Mongolia	S2A-S3Z	Bangladesh
D6A-D6Z	Comoros	JWA-JXZ	Norway	S5A-S5Z	Slovenia
D7A-D9Z	South Korea	JYA-JYZ	Jordan	S6A-S6Z	Singapore
EAA-EHZ	Spain	JZA-JZZ	Indonesia	S7A-S7Z	Seychelles
EIA-EJZ	Ireland	J2A-J2Z	Djibouti	S8A-S8Z	South Africa
EKA-EKZ	Armenia	J3A-J3Z	Grenada	S9A-S9Z	Sao Tome and Principe
ELA-ELZ	Liberia	J4A-J4Z	Greece		
EMA-EOZ	Ukraine	J5A-J5Z	Guinea-Bissau	TAA-TCZ	Turkey
EPA-EQZ	Iran	J6A-J6Z	Saint Lucia	TDA-TDZ	Guatemala
ERA-ERZ	Moldova	J7A-J7Z	Dominica	TEA-TEZ	Costa Rica
ESA-ESZ	Estonia	J8A-J8Z	St. Vincent and the Grenadines	TFA-TFZ	Iceland
ETA-ETZ	Ethiopia			TGA-TGZ	Guatemala
EUA-EWZ	Belarus	KAA-KZZ	United States of America	THA-THZ	France
				TIA-TIZ	Costa Rica

Call Sign Series	Allocated to	Call Sign Series	Allocated to	Call Sign Series	Allocated to
TJA-TJZ	Cameroon	WAA-WZZ	United States of America	3AA-3AZ	Monaco
TKA-TKZ	France			3BA-3BZ	Mauritius
TLA-TLZ	Central African Republic	XAA-XIZ	Mexico	3CA-3CZ	Equatorial Guinea
		XJA-XOZ	Canada		
TMA-TMZ	France	XPA-XPZ	Denmark	•3DA-3DM	Swaziland
TNA-TNZ	Congo	XQA-XRZ	Chile	•3DN-3DZ	Fiji
TOA-TQZ	France	XSA-XSZ	China	3EA-3FZ	Panama
TRA-TRZ	Gabon	XTA-XTZ	Burkina Faso	3GA-3GZ	Chile
TSA-TSZ	Tunisia	XUA-XUZ	Cambodia	3HA-3UZ	China
TTA-TTZ	Chad	XVA-XVZ	Viet Nam	3VA-3VZ	Tunisia
TUA-TUZ	Ivory Coast	XWA-XWZ	Laos	3WA-3WZ	Viet Nam
TVA-TXZ	France	XXA-XXZ	Portugal	3XA-3XZ	Guinea
TYA-TYZ	Benin	XYA-XZZ	Myanmar	3YA-3YZ	Norway
TZA-TZZ	Mali	YAA-YAZ	Afghanistan	3ZA-3ZZ	Poland
T2A-T2Z	Tuvalu	YBA-YHZ	Indonesia	4AA-4CZ	Mexico
T3A-T3Z	Kiribati	YIA-YIZ	Iraq	4DA-4IZ	Philippines
T4A-T4Z	Cuba	YJA-YJZ	Vanuatu	4JA-4KZ	Azerbaijan
T5A-T5Z	Somalia	YKA-YKZ	Syria	4LA-4LZ	Georgia
T6A-T6Z	Afghanistan	YLA-YLZ	Latvia	4MA-4MZ	Venezuela
T7A-T7Z	San Marino	YMA-YMZ	Turkey	4NA-4OZ	Yugoslavia
T8A-T8Z	Palau	YNA-YNZ	Nicaragua	4PA-4SZ	Sri Lanka
T9A-T9Z	Bosnia and Herzegovina	YOA-YRZ	Romania	4TA-4TZ	Peru
		YSA-YSZ	El Salvador	*4UA-4UZ	United Nations
UAA-UIZ	Russian Federation	YTA-YUZ	Yugoslavia	4VA-4VZ	Haiti
		YVA-YYZ	Venezuela	4WA-4WZ	UNTAET (E.Timor)
UJA-UMZ	Uzbekistan	YZA-YZZ	Yugoslavia		
UNA-UQZ	Kazakhstan	Y2A-Y9Z	Germany	4XA-4XZ	Israel
URA-UZZ	Ukraine	ZAA-ZAZ	Albania	*4YA-4YZ	International Civil Aviation Organization
VAA-VGZ	Canada	ZBA-ZJZ	United Kingdom of Great Britain and Northern Ireland		
VHA-VNZ	Australia				
VOA-VOZ	Canada			4ZA-4ZZ	Israel
VPA-VQZ	United Kingdom of Great Britain and Northern Ireland			5AA-5AZ	Libya
		ZKA-ZMZ	New Zealand	5BA-5BZ	Cyprus
		ZNA-ZOZ	United Kingdom of Great Britain and Northern Ireland	5CA-5GZ	Morocco
				5HA-5IZ	Tanzania
†VRA-VRZ	China (Hong Kong)			5JA-5KZ	Colombia
				5LA-5MZ	Liberia
VSA-VSZ	United Kingdom of Great Britain and Northern Ireland	ZPA-ZPZ	Paraguay	5NA-5OZ	Nigeria
		ZQA-ZQZ	United Kingdom of Great Britain and Northern Ireland	5PA-5QZ	Denmark
				5RA-5SZ	Madagascar
				5TA-5TZ	Mauritania
VTA-VWZ	India			5UA-5UZ	Niger
VXA-VYZ	Canada	ZRA-ZUZ	South Africa	5VA-5VZ	Togo
VZA-VZZ	Australia	ZVA-ZZZ	Brazil	5WA-5WZ	Western Samoa
V2A-V2Z	Antigua and Barbuda	Z2A-Z2Z	Zimbabwe	5XA-5XZ	Uganda
		Z3A-Z3Z	Macedonia (Former Yugoslav Republic)	5YA-5ZZ	Kenya
V3A-V3Z	Belize			6AA-6BZ	Egypt
V4A-V4Z	Saint Kitts and Nevis			6CA-6CZ	Syria
				6DA-6JZ	Mexico
V5A-V5Z	Namibia	2AA-2ZZ	United Kingdom of Great Britain and Northern Ireland	6KA-6NZ	South Korea
V6A-V6Z	Micronesia			6OA-6OZ	Somalia
V7A-V7Z	Marshall Islands			6PA-6SZ	Pakistan
V8A-V8Z	Brunei			6TA-6UZ	Sudan

6VA-6WZ	Senegal	8OA-8OZ	Botswana	9MA-9MZ	Malaysia
6XA-6XZ	Madagascar	8PA-8PZ	Barbados	9NA-9NZ	Nepal
6YA-6YZ	Jamaica	8QA-8QZ	Maldives	9OA-9TZ	Congo Rep
6ZA-6ZZ	Liberia	8RA-8RZ	Guyana	9UA-9UZ	Burundi
7AA-7IZ	Indonesia	8SA-8SZ	Sweden	9VA-9VZ	Singapore
7JA-7NZ	Japan	8TA-8YZ	India	9WA-9WZ	Malaysia
7OA-7OZ	Yemen	8ZA-8ZZ	Saudi Arabia	9XA-9XZ	Rwanda
7PA-7PZ	Lesotho	9AA-9AZ	Croatia	9YZ-9ZZ	Trinidad and
7QA-7QZ	Malawi	9BA-9DZ	Iran		Tobago
7RA-7RZ	Algeria	9EA-9FZ	Ethiopia		
7SA-7SZ	Sweden	9GA-9GZ	Ghana		
7TA-7YZ	Algeria	9HA-9HZ	Malta		
7ZA-7ZZ	Saudi Arabia	9IA-9JZ	Zambia		
8AA-8IZ	Indonesia	9KA-9KZ	Kuwait		
8JA-8NZ	Japan	9LA-9LZ	Sierra Leone		

• Half series
* Series allocated to an international organization
†Provisional allocation in accordance with S19.33

Index

Notes

Notes

Notes

Notes

Notes

FEEDBACK

Please use this form to give us your comments on this book and what you'd like to see in future editions, or e-mail us at **pubsfdbk@arrl.org** (publications feedback). If you use e-mail, please include your name, call, e-mail address and the book title, edition and printing in the body of your message. Also indicate whether or not you are an ARRL member.

Where did you purchase this book?
□ From ARRL directly   □ From an ARRL dealer

Is there a dealer who carries ARRL publications within:
□ 5 miles   □ 15 miles   □ 30 miles   of your location?   □ Not sure.

License class:
□ Novice   □ Technician   □ Technician Plus   □ General   □ Advanced   □ Extra

Name _____

Daytime Phone (   ) _____

Address _____

City, State/Province, ZIP/Postal Code _____

If licensed, how long? _____

Other hobbies _____

_____

Occupation _____

ARRL member?   □ Yes □ No

Call Sign _____

Age _____

E-mail _____

For ARRL use only	ON THE AIR
Edition	1 2 3 4 5 6 7 8 9 10 11 12
Printing	1 2 3 4 5 6 7 8 9 10 11 12

From _____

_____

_____

EDITOR, ON THE AIR WITH HAM RADIO
ARRL
225 MAIN STREET
NEWINGTON CT 06111-1494

— — — — — — — — — — please fold and tape — — — — — — — — — — — —